Agroforestry: A Comprehensive Introduction

Agroforestry: A Comprehensive Introduction

Morgan Clayton

STATES
ACADEMIC PRESS
www.statesacademicpress.com

Agroforestry: A Comprehensive Introduction
Morgan Clayton
ISBN: 978-1-63989-036-1 (Hardback)

STATES
ACADEMIC PRESS

Published by States Academic Press,
109 South 5th Street,
Brooklyn, NY 11249, USA

Cataloging-in-Publication Data

Agroforestry : a comprehensive introduction / Morgan Clayton.
p. cm.
Includes bibliographical references and index.
ISBN 978-1-63989-036-1
1. Agroforestry. 2. Agroforestry systems. 3. Agriculture. 4. Forests and forestry. I. Clayton, Morgan.
S494.5.A45 A37 2022
634.99--dc23

For more information regarding States Academic Press and its products, please visit the publisher's website www.statesacademicpress.com

TABLE OF CONTENTS

TABLE OF CONTENTS

PREFACE

It is with great pleasure that I present this book. It has been carefully written after numerous discussions with my peers and other practitioners of the field. I would like to take this opportunity to thank my family and friends who have been extremely supporting at every step in my life.

Agroforestry is the system that is used to monitor land usage. It involves planting different shrubs and trees around and among pastureland and crops. This field has evolved with the integration of agriculture and forestry. The systems of agroforestry are used to enhance biodiversity, reduce soil erosion, enhance wildlife habitats, contribute towards sustainable agricultural systems and improve adaptation to climate change. The field is based on the theories of agroecology and ecology. Tropical agroforestry and temperate agroforestry are some of the areas where this field finds its application. Some types of systems used in this field include crop-over-tree systems, shade crops, itteri agroforestry, and alley and strip cropping. This book is compiled in such a manner, that it will provide an in-depth knowledge about the theory and practice of agroforestry. It also unfolds the innovative aspects related to this discipline which will be crucial for the holistic understanding of the subject matter. This book, with its detailed analyzes and data, will prove immensely beneficial to professionals and students involved in this area at various levels.

The chapters below are organized to facilitate a comprehensive understanding of the subject:

Chapter – Introduction

Forestry is the art and practice of managing, conserving and studying forests and plantations. Agroforestry is the combination of agriculture and forestry. It is a land-use system in which shrubs and trees are grown around pastureland to increase biodiversity and reduce erosion. This is an introductory chapter which will briefly introduce about agroforestry.

Chapter – Classification of Agroforestry Systems

Agroforestry can be classified into a number of systems such as silviculture, agri-silviculture system, silvopasture, dehesa, agri-silve-pastoral system, forest gardening, etc. The topics elaborated in this chapter will help in gaining a better perspective of the classification of agroforestry systems.

Chapter – Silviculture: An Integrated Study

The art, science and practice of controlling the structure, growth and quality of forests for meeting the needs of timber production is termed as silviculture. Coppice method, clearcut method, ecological thinning, selection cutting, seed tree method, etc. fall under its domain. This chapter has been carefully written to provide an easy understanding of the related concepts of silviculture.

Chapter – Sustainable Forest Management

Forest management deals with the economic, social, technical, environmental and ecological aspects of forest. Sustainable forest management maintains balance between all these aspects by following the principles of sustainable management. This chapter discusses the subject of sustainable forest management in detail.

Chapter – Diverse Aspects of Agroforestry

There are diverse aspects that are studied under agroforestry including modeling agroforestry systems, bamboo based agroforestry and agroforestry watershed management. This chapter sheds light on these diverse aspects of agroforestry for an in-depth understanding of the subject.

Morgan Clayton

Introduction

CHAPTER 1

Forestry is the art and practice of managing, conserving and studying forests and plantations. Agroforestry is the combination of agriculture and forestry. It is a land-use system in which shrubs and trees are grown around pastureland to increase biodiversity and reduce erosion. This is an introductory chapter which will briefly introduce about agroforestry.

FORESTS

A forest is a type of ecosystem in which there is high density of trees occupying a relatively large area of land. An ecosystem is an ecological unit consisting of a biotic community (an assemblage of plant, animal, and other living organisms) together with its abiotic environment (such as soil, rocks, water, temperature, slope of the land, etc.). In the case of a forest, trees dominate the biotic landscape, although there are also other plants and animals. There are many types of forest, such as rainforests and temperate hardwood forest.

There are many definitions of a forest, based on various criteria. These plant communities cover large areas of the globe and function as habitats for organisms, hydrologic flow modulators, and soil conservers, constituting one of the most important aspects of the Earth's biosphere.

"Forest" meant an uncultivated area legally set aside for hunting by feudal nobility, and these hunting forests were not necessarily wooded much, if at all. However, as hunting forests did often include considerable areas of woodland, the word forest eventually came to mean wooded land more generally.

Forests provide innumerable values to people, provide aspects that address both physical needs (shelter, food, etc.) as well as the internal nature of people (beauty, diversity of animals and plants, recreation, etc.). Some of the values of forests can be converted to commercial value, with trees providing timber for construction and housing, paper, firewood, and so forth. The selling of timber allows an economic valuation to be established. However, for many of the benefits offered by forests, it would be difficult to compute the economic value. These benefits include climate control, ecological values, water retention, and aesthetic, peace of mind, and recreational values.

Forests can be found in all regions capable of sustaining tree growth, at altitudes up to the tree line, except where natural fire frequency is too high, or where the environment has been impaired by natural processes or by human activities. The tree line or timberline is the edge of the habitat at which trees are capable of growing; beyond the tree line, they are unable to grow due to inappropriate environmental conditions.

A deciduous broadleaf (Beech) forest in Slovenia.

As a general rule, forests dominated by angiosperms (broadleaf forests) are more species-rich than those dominated by gymnosperms (conifer, montane, or needleleaf forests), although exceptions exist (for example, species-poor aspen and birch stands in northern latitudes).

A forest on San Juan Island in Washington.

Forests sometimes contain many tree species within a small area (as in tropical rainforests and temperate deciduous forests), or relatively few species over large areas (e.g., taiga and arid montane coniferous forests). Forests are often home to many animal and plant species, and biomass per unit area is high compared to other vegetation communities. Much of this biomass occurs below-ground in the root systems and as partially decomposed plant detritus. The woody component of a forest contains lignin, which is relatively slow to decompose compared with other organic materials such as cellulose or carbohydrate.

Maple and Oak (broadleaf, deciduous) forest in Wisconsin during winter.

Forests are differentiated from woodlands by the extent of canopy coverage: in a forest the branches and foliage of separate trees often meet or interlock, although there can be gaps of varying sizes within an area referred to as forest. A woodland has a more continuously open canopy, with trees spaced further apart, which allows more sunlight to penetrate to the ground between them.

Among the major forested biomes are:

- Rainforest (tropical and temperate).
- Taiga.
- Temperate hardwood forest.
- Tropical dry forest.

FOREST TYPES AND CLASSIFICATION

Forests have been classied in different ways. They have been classied according to the Biomes in which they exist, combined with leaf longevity of the dominant Species i.e. whether they are evergreen or deciduous. Another Classication is classied whether the Forests are composed predominantly of BroadLeaf Trees, Coniferous (needle-leaved) Trees or Mixed.

- UNEP-WCMC Classication United Nations Environment Programme - World Conservation Monitoring Centre divides the World's Forest into 26 major types, which reect climatic zones as well as the principal types of Trees. These 26 Forest categories are used to enable the translation of Forest types from national and regional classication systems to a harmonised global one.

Temperate and Boreal Forest Types

- Evergreen Needleleaf Forest: Natural Forest with > 30% canopy cover, in which the canopy is predominantly (> 75%) Needleleaf and Evergreen.
- Deciduous Needleleaf Forest: Natural forest with >30% canopy cover, in which the canopy is predominantly (>75%) Needleleaf and Deciduous.
- Mixed Broadleaf/Needleleaf Forest: Natural Forest with > 30% canopy cover, in which the canopy is composed of a more or less even mixture of Needleleaf and Broadleaf crowns (between 50:50% and 25:75%).
- Broad Leaf Evergreen Forest: Natural Forests with > 30% canopy cover, the canopy being > 75% evergreen and Broadleaf.
- Deciduous Broadleaf Forest: Natural Forests with > 30% canopy cover, in which > 75% of the canopy is deciduous and Broadleaves predominate (> 75% of canopy cover).
- Freshwater Swamp Forest: Natural Forests with > 30% canopy cover, composed of Trees with any mixture of leaf type and seasonality, but in which the predominant environmental characteristic is a water logged soil.
- Sclerophyllous Dry Forest: Natural Forest with > 30% canopy cover, in which the canopy is mainly composed of sclerophyllous Broadleaves and is > 75% evergreen.
- Disturbed Natural Forest: Any Forest type above that has in its interior significant areas of disturbance by people, including clearing, felling for wood extraction, anthropogenic fires, road construction, etc.
- Sparse Trees and Parkland: Natural Forests in which the Tree canopy cover is between 10-30%, such as in the steppe regions of the world. Trees of any type (e.g., needleleaf, broadleaf, palms).
- Exotic Species Plantation: Intensively managed Forests with > 30% canopy cover,which have been planted by people with Species not naturally occurring in that country.
- Native Species Plantation: Intensively managed Forests with > 30% canopy cover,which have been planted by people with Species that occur naturally in that country.

Tropical Forest Types

- Lowland Evergreen Broadleaf Rain Forest: Natural Forests with > 30% canopy cover, below 1200m altitude that display little or no seasonality, the canopy being >75% evergreen Broadleaf.

- Lower Montane Forest: Natural Forests with > 30% canopy cover, between 1200-1800m altitude, with any seasonality regime and leaf type mixture.
- Upper Montane Forest: Natural Forests with > 30% canopy cover, above 1800m altitude, with any seasonality regime and leaf type mixture.
- Freshwater Swamp Forest: Natural Forests with > 30% canopy cover, below 1200m altitude, composed of Trees with any mixture of leaf type and seasonality, but in which the predominant environmental characteristic is a waterlogged soil.
- Semi-evergreen Moist Broadleaf Forest: Natural Forests with > 30% canopy cover, below 1200m altitude in which between 50-75% of the canopy is evergreen, > 75% are Broadleaves, and the Trees display seasonality of fowering and fruiting.
- Mixed Broadleaf/Needleaf Forest: Natural Forests with >30% canopy cover, below 1200m, in which the canopy is composed of a more or less even mixture of Needleleaf and Broadleaf crowns(between 50:50% and 25:75%).
- Needleleaf Forest: Natural Forest with > 30% canopy cover, below 1200m altitude, in which the canopy is predominantly (> 75%) Needleleaf.
- Mangroves: Natural Forests with > 30% canopy cover, composed of Species of mangrove Tree, generally along coasts in or near brackish or salt water.
- Disturbed Natural Forest: Any Forest type above that has in its interior signi_ cant areas of disturbance by people, including clearing, felling for wood extraction, anthropogenic _res, road construction, etc.
- Deciduous/Semi-Deciduous Broadleaf Forest: Natural Forests with > 30% canopy cover, below 1200m altitude in which between 50-100% of the canopy is Deciduous and Broadleaves predominate (> 75% of canopy cover).
- Sclerophyllous Dry Forest: Natural Forests with > 30% canopy cover, below 1200m altitude, in which the canopy is mainly composed of sclerophyllous Broadleaves and is > 75% evergreen.
- Thorn Forest: Natural Forests with > 30% canopy cover, below 1200m altitude, in which the canopy is mainly composed of deciduous Trees with thorns and succulent phanerophytes with thorns may be frequent.
- Sparse Trees and Parkland: Natural Forests in which the Tree canopy cover is between 10-30%, such as in the savannah regions of the world. Trees of any type (e.g., Needleleaf, Broadleaf, Palms).
- Exotic Species Plantation: Intensively managed Forests with > 30% canopy cover, which have been planted by people with Species not naturally occurring in that country.

- Native Species Plantation: Intensively managed Forests with > 30% canopy cover, which have been planted by people with Species that occur naturally in that country.

These 26 major types can be reclassifed into 6 broader categories:

- Temperate Needleleaf: Temperate Needleleaf Forests cover a larger area of the world than any other Forest types. They mostly occupy the higher latitude regions of the northern hemisphere, as well as high altitude zones and some warm temperate areas, especially on nutrient-poor or otherwise unfavourable soils. These Forests are composed entirely, or nearly so, of coniferous Species (Coniferophyta). In the Northern Hemisphere pines Pinus, spruces Picea, larches Larix, silver firs Abies, Douglas firs Pseudotsugaand hemlocks Tsuga, make up the canopy, but other taxa are also important. In the southern hemisphere most coniferous Trees, members of the Araucariaceae and Podocarpaceae, occur in mixtures with Broadleaf Species that are classed as Broadleaf and Mixed Forests.
- Temperate Broadleaf and Mixed: Temperate Broadleaf and mixed Forests, which include a substantial component of Trees in the Anthophyta, cover over 6.5 million km2 of the Earth's surface. They are generally characteristic of the warmer temperate latitudes, but extend to cool temperate ones, particularly in the southern hemisphere. They include such Forest types as the mixed deciduous Forests of the USA and their counterparts in China and Japan, the Broadleaf evergreen rain Forests of Japan, Chile and Tasmania, the sclerophyllous Forests of Australia, the Mediterranean and California, and the southern beech Nothofagus Forests of Chile and New Zealand.
- Tropical Moist: Tropical moist Forests cover more than 11 million km^2 of the humid tropics and include many different Forest types. The best known and most extensive are the Lowland Evergreen Broadleaf Rainforests, which make up over half this area and include, for example: the seasonally inundated varzea and igapó forests and the terra rme Forests of the Amazon Basin; the Peat Forests and moist dipterocarp Forests of Southeast Asia; and the high Forests of the Congo Basin. The Forests of tropical mountains are also included in this broad category, generally divided into upper and lower montane formations on the basis of their physiognomy, which varies with altitude. The montane Forests include cloud Forest, those Forests at middle to high altitude, which derive a signicant part of their water budget from cloud, and support a rich abundance of Vascular and Non-Vascular Epiphytes. Mangrove Forests also fall within this broad category, as do most of the Tropical Coniferous Forests of Central America.
- Tropical Dry: Tropical dry Forests are characteristic of areas in the tropics affected by seasonal drought. Such seasonal climates characterise much of the tropics, but less than 4 million km^2 of tropical dry Forests remain. The seasonality of rainfall is usually reected in the deciduousness of the Forest canopy,

with most Trees being leaess for several months of the year. However, under some conditions, e.g. less fertile soils or less predictable drought regimes, the proportion of evergreen Species increases and the Forests are characterised as "sclerophyllous". Thorn Forest, a dense Forest of low stature with a high frequency of thorny or spiny Species, is found where drought is prolonged, and especially where grazing animals are plentiful. On very poor soils, and especially where re is a recurrent phenomenon, Woody Savannas develop.

- Sparse Trees and Parkland: Sparse Trees and parkland are Forests with open canopies of 10-30% crown cover. They occur principally in areas of transition from Forested to non-Forested landscapes. The two major zones in which these ecosystems occur are in the boreal region and in the seasonally dry tropics. At high latitudes, north of the main zone of boreal Forest or Taiga, growing conditions are not adequate to maintain a continuous closed Forest cover, so Tree cover is both sparse and discontinuous. This vegetation is variously called open Taiga, open lichen woodland, and Forest tundra. It is Species-poor, has high bryophyte cover and is frequently affected by fire.

- Forest Plantations: Forest plantations, generally intended for the production of timber and pulpwood increase the total area of Forest worldwide. In 1999 FAO has estimated that total plantation area in developed countries is about 600,000 km^2 and in developing countries it is about 550,000 km^2. Commonly mono-specic and composed of introduced Tree Species, these ecosystems are not generally important as habitat for native biodiversity. However, they can be managed in ways that enhance their biodiversity protection functions and they are important providers of ecosystem services such as maintaining nutrient capital, protecting watersheds and soil structure as well as storing carbon. They may also play an important role in alleviating pressure on natural Forests for timber and fuel-wood production.

WWF (World Wide Fund for Nature) Classification

The type of Forest depends mainly on location - that is, distance from equator and altitude - and climate.

Basic Classification

- Tropical Forest: These type of Forests are characterised by their location neat the equator. They have year-round high temperatures and abundant rainfall which makes them a dense and lush with vegetation. They are vital storehouses of biodiversity on the planet.

- Sub-tropical Forests: These are found to the south and north of the Tropical Forests. The Trees here are adapted to resist the summer drought.

- Mediterranean Forests: These Forests are found to the south of the temperate regions around the coasts of the Mediterranean, California, Chile and Western Australia. The growing season is short ans almost all Trees are evergreen, but mixed hardwood and softwood.
- Temperate Forests: These Forest are found in eastern North America, north-eastern Asia, and western and eastern Europe. Temperate Forests are a mix of Deciduous and Coniferous Evergreen Trees. Usually, the Broad-Leaved Hardwood Trees shed leaves annually. There are well-dened seasons with a distinct winter and sufcient rainfall.
- Coniferous Forests: Coniferous Forests are found in the cold, windy regions around the poles. There are both Hard-woods and Conifers found in this region. The Conifers are evergreen and structurally adapted to withstand the long drought-like conditions of the long winters, whereas the Hard-Woods are Deciduous.
- Montane Forests: These Forests receive most of their precipitation from the mist or fog that comes up from the lowlands and as such are also known as cloud Forests. Some of these Montane Woodlands and Grasslands are found in high-elevation tropical, subtropical and temperate zones. Plants and animals in these Forests are adapted to withstanding the cold, wet conditions and intense sunlight. Trees are mainly Conifers.

Tropical and Subtropical Moist Broadleaf Forest

These types of Forests are generally found in large, discontinuous patches centred on the equatorial belt and between the Tropics of Cancer and Capricorn. These Forests are found around the world, particularly in the Indo-Malayan Archipelagos, the Amazon Basin, and the African Congo. In general, biodiversity is highest in the Forest canopy which can be divided into five layers: Overstory canopy with emergent crowns, a medium layer of canopy, lower canopy, shrub level, and nally understory.

- These Forests are characterized by low variability in annual temperature and high levels of rainfall (>200 centimetre annually).
- Forest composition is dominated by semi-evergreen and evergreen deciduous Tree Species. These Trees number in thousands and contribute to the highest levels of Species diversity in any terrestrial major Habitat type.
- A perpetually warm, wet climate promotes more explosive plant growth than in any other environment on Earth. A Tree here may grow over 75 feet in height in just 5 years.

- The canopy is home to many of the Forest's animals, including Apes, Monkeys, Hornbills, Toucans, and the Harpy Eagle. Below the canopy, a lower understory hosts to Snakes and Big Cats. The Forest oor, relatively clear of undergrowth due to the thick canopy above, is prowled by other Animals such as Gorillas and Deer.

Some of the Ecoregions of Tropical and Subtropical Moist Broadleaf Forest:

- Seychelles and Mascarenes Moist Forests in Islands to the north and east of Madagascar, off the east coast of Africa.
- East African Coastal Forests in Kenya, Somalia, Tanzania.
- Albertine Rift Montane Forests in Burundi, Democratic Republic of Congo, Rwanda, Tanzania, Uganda.
- Sulawesi Moist Forests in Sulawesi, a large island to the south-east of Borneo, in Indonesia.
- New Guinea Montane Forests in The island of New Guinea, north of Australia.
- Queensland Tropical Forests in Northeastern Australia.
- Lord Howe-Norfolk Islands Forests in Islands off the east coast of Australia in the Tasman Sea.
- South-western Ghats Moist Forests in Western coast of India.
- Northern Indochina Subtropical Moist Forests in Southern China, northern Laos, Myanmar, Thailand, and Vietnam.
- Talwan Montane Foorests in Talwan, a large island off the eastern coast of China.
- Peninsular Malaysian Lowland and Mountain Forests in Malaysia, Singapore, Thailand.
- Nansei Shoto Archipelago Forests between southern Japan and Taiwan.
- Eastern Deccan Plateau Moist Forests in India.
- Cardamom Mountains Moist Forests in Cambodia and Thailand.
- Talamancan and Isthmian Pacic Forests in Costa Rica and western Panama.
- Central Andean Yungas in Argentina, Bolivia, and Peru.
- Atlantic Forests in Argentina, Brazil, Paraguay.
- South Pacic Islands Forests in American Samoa, Cook Islands, Fiji, French Polynesia, Niue, Samoa, Tonga, Wallis and Futuna Islands.

Tropical and Subtropical Dry Broadleaf Forest

These are found in southern Mexico, south-eastern Africa, the Lesser Sundas, central India, Indochina, Madagascar, New Caledonia, eastern Bolivia and central Brazil, the Caribbean, valleys of the northern Andes, and along the coasts of Ecuador and Peru.

- These Forest are are warm year-round, and may receive several hundred centimetres or rain per year but they also experience long dry seasons which last several months and vary with geographic location.
- Deciduous Trees predominate these Forests, and during the drought a leaess period occurs, which varies with Species type.
- The newly bare Trees open up the canopy layer, enabling sunlight to reach ground level and facilitate the growth of thick underbrush.
- Tropical Dry Forests are home to a wide variety of wildlife including monkeys, large cats, parrots, various rodents, and ground dwelling birds. Many of these Species are adapted to the prevalent environment.

Some of the Ecoregions of Tropical and Subtropical Dry Broadleaf Forest:

- Madagascar Dry Forests in Western coast of Madagascar.
- Nusu Tenggara Dry Forests in South-east Asia: Indonesia.
- New Caledonia Dry Forests in New Caledonia, a large island north-east of Australia.
- Indochina Dry Forests in Cambodia, Laos, Thailand, Vietnam.
- Chhota-Nagpur Dry Forests in Eastern India.
- Mexican Dry Forest in Guatemala and Mexico.
- Tumbesian-Andean Valleys Dry Forests in Colombia, Ecuador, and Peru.
- Chiquitano Dry Forest in Bolivia and Brazil.
- Atlantic Dry Forests in North-eastern Brazil.
- Hawaii's Dry Forests in Hawaiian Islands.

Tropical and Sub-tropical Coniferous Forest

These are found predominantly in North and Central America, these tropical regions experience low levels of precipitation and moderate variability in temperature.

- Tropical and Subtropical Coniferous Forests are characterized by diverse Species of conifers, whose needles are adapted to deal with the variable climatic conditions.

- These biomes feature a thick, closed canopy which blocks light to the oor and allows little underbrush. As a result, the ground is often covered with fungi and ferns. Shrubs and small Trees compose a diverse understory.

Some of the Ecoregions of Tropical and Suptropical Coniferous Forest:

- Sierra Madre Oriental and Occidental Pine-Oak Forests in Southwestern United States and Central Mexico.
- Greater Antillean Pine Forests in Cuba, Dominican Republic, Haiti, and Greater Antilles Islands.
- Mesoamerican Pine-oak Forests in Mexico, Guatemala, Nicaragua, Honduras, El Salvador.

Temperate Broadleaf and Mixed Forest

These are found in central China, eastern North America, Caucasus, the Himalayas, southern Europe, and the Russian Far East.

- Forests here experience a wide range of variability in temperature and precipitation.
- In regions where rainfall is broadly distributed throughout the year, Deciduous Trees mix with Species of evergreens. Species such as Oak (Quercus spp.), Beech (Fagus spp.), Birch (Betupa spp.) and Maple (Acer spp.) ll up the landscape of these Forests.
- Structurally, these Forests are characterized by 4 layers: a canopy composed of mature full-sized dominant Species and a slightly lower layer of mature Trees, a shrub layer, and understory layer of Grasses and other Herbaceous Plants.
- Most biodiversity in these Forests is concentrated much closer to the Forest floor.

Some of the Ecoregions of Temperate Broadleaf and Mixed Forest:

- Eastern Australia Temperate Forests in South-eastern Australia.
- Tasmanian Temperate Rain Forests in Tasmania - an island off South-eastern Australia.
- New Zealand Temperate Forests in Southern island of New Zealand.
- Eastern Himalayan Broadleaf and Conifer Forests in Bhutan, China, India, Myanmar, Nepal.
- Western Himalayan Temperate Forests in in India, Nepal, Pakistan.

- Appalachian and Mixed Mesophytic Forests in Eastern United States.
- Russian Far East Temperate Forests in South-eastern Pacic coast of Russia.

Temperate Coniferous Forest

Temperate Coniferous Forest are found predominantly in areas with warm summers and cool winters, and vary enormously in their kinds of plant life. These include the Pacic Northwest, the Validivian Forests of South-western South America, the Rainforests of New Zealand and Tasmania, the North-eastern Atlantic (small, isolated pockets in Ireland, Scotland, and Iceland), southwestern Japan, and those of the eastern Black Sea .In some, Needleleaf Trees dominate, while others are home primarily to Broadleaf Evergreen Trees or a mix of both Tree types.

- These are common in the coastal areas of regions that have mild winters and heavy rainfall, or inland in drier climates or Montane areas.
- Many Species of Trees inhabit these Forests including Pine, Cedar, Fir, and Redwood which are huge Trees, especially the Redwood.
- The understory (vegetation which grows under the tall Trees) also contains a wide variety of Herbaceous and Shrub Species.
- Structurally, these Forests are rather simple, consisting of 2 layers generally: an overstory and understory. However, some Forests may support a layer of shrubs. Pine Forests support an Herbaceous ground layer that may be dominated by grasses and forbs that lend themselves to ecologically important wildres. In contrast, the moist conditions found in temperate rain Forests favour the dominance by ferns and some forbs.

Some of the Ecoregions of Temperate Coniferous Forest:

- Pacic Temperate Rainforests in West Coast of United States and Canada.
- Klamath-Siskiyou Coniferous Forests in California and Oregon.
- Sierra Nevada Coniferous Forests in Western United States.
- South-eastern Coniferous and Broadleaf Forests in South-eastern United States.
- Valdivian Temperate Rainforests / Juan Fernandez Islands in Argentina, Chile.
- European-Mediterranean Montane Mixed Forests in South-eastern Europe.
- Caucasus-Anatolian-Hyrcanian Temperate Forests in Armenia, Azerbaijan, Bulgaria, Georgia, Iran, Russia, Turkey, Ukraine.
- Altai-Sayan Montane Forests in China, Kazakhstan, Mongolia, Russia.

- Hengduan Shan Coniferous Forests in China.

Boreal Forests/Taiga

- These Forests are found in inland Canada, Alaska, Sweden, Finland, inland Norway, much of Russia (especially Siberia), as well as parts of the extreme northern continental United States (northern Minnesota, Michigan's Upper Peninsula, northern Wisconsin, Up-state New York, Vermont, New Hampshire, and Maine), northern Kazakhstan, northern Mongolia, and northern Japan (Hokkaido).
- There are two major types of Taiga:
    - The Southern Part is the Closed Canopy Forest, consisting of many closely-spaced Trees with mossy ground cover. In clearings in the Forest, shrubs and wild owers are common, such as the Fireweed.
    - Lichen Woodland or Sparse Taiga, with Trees that are farther-spaced and lichen ground cover; the latter is common in the northernmost Taiga.
- The Taiga or Boreal Forest has a subarctic continental climate with very large temperature range between seasons, but the long and cold winter is the dominant feature. The short summer (24-hr average 10 °C or more) lasts 1–3 months and always less than 4 months. The winters last 5 – 7 months, with average temperatures below freezing. Temperatures vary from -54 °C to 30 °C (-65 °F to 86 °F) throughout the whole year.
- The Taiga experiences relatively low precipitation throughout the year (generally 200–750 mm annually, 1,000 mm in some areas), primarily as rain during the summer months, but also as fog and snow. As evaporation is also low for most of the year, precipitation exceeds evaporation, and is sufcient to sustain the dense vegetation growth. Snow may remain on the ground for as long as nine months in the northernmost extensions of the Taiga ecozone.
- The Forests of the Taiga are largely coniferous, dominated by Larch, Spruce, Fir, and Pine.
- The Taiga is home to a number of various Mammals, such as Moose, Reindeer, American Black Bear, Grizzly Bear and Wolverine.

Forests are found in all regions of the World which are capable of sustaining Tree growth, at altitudes up to the Tree line, except where natural re frequency or other disturbance is too high or hostile environment like Antarctica.

General Characteristics of Forests

- Domination of Trees: Forests are mostly dominated by large Plants like Trees. Some Trees especially in Rainforests gain a lot of height to get more sunlight.

Most of other Plants face tough competition from large Plants as these tend to receive most of the nutrition from the environment. Some Plants have evolved to become Creeper and Climbers to combat this competition. They climb over Trees to reach the sunlight as well.

- Forest Canopy: Forest Canopy refers to the cover of Tall Forests. This top portion of a community of Trees or Plant Crowns serves as the interface between the atmosphere and the land. The Canopy is also the upper habitat for other biological organisms in a Forest ecosystem. The structure of Forest Canopy varies from Forest to Forest because because of the availability of nutrients, Tree arrangement and differences in biological Species. The Forest Canopy is an Ecosystem in itself as it supports a variety of life which is not found in the Forest itself.
- Availability of good amount of Water: Most of the Forests receive sufcient rains to support the organisms which depend on it. It is only in dry season that these may face shortage of water. This rain water forms ponds and streams which supports further life forms like Plant and Animals.
- Great Biodiversity: Forests are one of the major Habitats which support a great biodiversity of life than any other Habitat. This is attributed to the easy availability of food and shelter. Scientists estimate that more than half of all the world's plant and animal Species live in tropical Rainforests. A part of this diversity are the local tribal people who depend on the Forest for almost all their needs.
- Forest Floor: The Forest Floor is composed of fallen leaves, stems, twigs, branches and bark on the surface of the soil. A Forest oor also contains organic and inorganic substances. The Forest Floor is inhabited by various living organisms, such as the fungi, bacteria and other microorganisms as it is rich in nutrients and mineral contents. The Forest Floor has a signicant role in the transfer of nutrients in the life cycle of the Forest ecosystem. The leaves that fall on the Forest Floor keep on piling up. These leaves decompose over a period of time and provide essential nutrients which promote the growth of Trees.
- Variation in Soil Fertility: The soil of Temperate Forests is fertile because Trees' leaves drop to the ground every fall. This litter contributes to the layers of organic material found in Forest soil. The old leaves become a source of food for bacteria and fungi. These organisms facilitate the breaking down of the leaves and other organic material. Decomposition enriches the Forest soil as it provides more nutrients to the living Trees and plants in the Forest ecosystem. However, the soil in tropical rain Forests has poor quality because of the torrential rains. The constant rain erodes and dissolves soil nutrients before the Trees can benet from them.
- Complex Ecosystems: Forests are among the most complex ecosystems in the World. Conifer Forests have the simplest structure: a Tree layer rising to about

98 feet (30 m), a shrub layer that is spotty or even absent, and a ground layer covered with lichens, mosses, and liverworts. Deciduous Forests are more complex; the Tree canopy is divided into an upper and lower story, while Rain Forest canopies are divided into at least three strata. The Forest oor in both of these Forests consists of a layer of organic matter overlying mineral soil. The humus layer of tropical soils is affected by the high levels of heat and humidity, which quickly decompose whatever organic matter exists. Fungi on the soil surface play an important role in the availability and distribution of nutrients, particularly in the northern coniferous Forests. Some Species of fungi live in partnership with the Tree roots, while others are parasitically destructive.

PHYSICAL ENVIRONMENT WITHIN FORESTS

Precipitation

All plant growth depends on the availability of water, which in forest ecosystems is provided primarily by precipitation, but can also be provided by access to groundwater or, in riparian forest communities, to stream water. Forest canopies intercept some precipitation, and thus only 50-80% of rainfall reaches the forest floor, of which some of that evaporates before entering the soil. Of the precipitation that reaches the soil, approximately 50-80% will be transpired back into the atmosphere by the vegetation. Thus, forests serve as large water pumps between the soil and the atmosphere.

Soil

Trees are rooted in soil, which provides nutrients and water that support tree growth. The availability of nutrients and water are dependent on soil texture (sand, silt, loam, clay content), pH, cation exchange capacity, and water holding capacity, that are determined not only by soil texture but also by organic matter content, and porosity. These properties are largely a function of the geology of a specific site, which is determined by the bedrock quality, weathered bedrock material, and organic matter incorporated into the weathered bedrock material. Depending on the soil properties, more or less nutrients or water become available to the trees. Nutrients are cycled between the biosphere and the pedosphere (soil) through tree uptake, litterfall, and decomposition. Some nutrients are re-absorbed, for example nitrogen, before the leaves are dropped. However, most nutrients are recycled into the soil via decomposition through microbes and fungi. More importantly, trees are the primary pathway by which water moves between the soil and the atmosphere. However, trees can also redistribute water within the soil. When trees have access to groundwater and the topsoil is dry, then hydraulic lift can occur. Hydraulic lift is the biologically mediated process of water redistribution through primarily plant roots, but can also be facilitated through, for example, mycorrhizal mycelia from lower moist soil layers to upper dryer soil layers. Recently, it has been also shown that roots transport water laterally in the soil to dryer soil regions.

Light

Forests alter microclimatological conditions due to their height, leaf area, and density, through the absorption of radiation and cooling through transpiration. For example, in a deciduous forest canopy, light reaches the floor in the winter time, but less so in the summer time when leaves are blocking the passage of light to the ground. This has consequences for photosynthesis, transpiration, and understory processes. Photosynthesis requires light and carbon dioxide (CO_2) which enters through stomata in the leaves and needles of forest tree species. Stomata are pores that can open and close to regulate the uptake of CO_2 and the loss of water (transpiration). This is a basic trade-off that all plants need to make, and highlights the critical role of stomata in controlling transpiration and photosynthesis. As light penetrates the canopy, it is attenuated (i.e., decreased). Thus leaves lower in the canopy receive less light and have to develop in such a way to maintain a positive carbon balance via photosynthesis. This is accomplished by orienting the leaves horizontally, and making them larger and thinner, which allows them to capture more light per amount of leaf material-in essence displaying shade leaf properties. Concomitant with reduced light, carboxylation efficiency, or the efficiency of carbon uptake in leaves, is also reduced. Leaf shape and placement, though, can play an important role. In coniferous canopies, due to the shoot structure and narrow leaf shape, more light penetrates, thus enhancing photosynthesis lower in the canopy and in the understory species. The effect of light bouncing in-between needles on a shoot, is referred to as the penumbral effect.

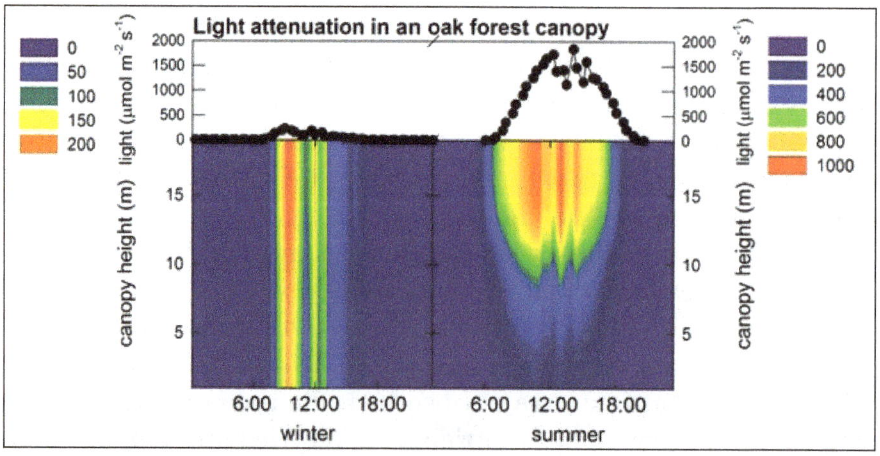

Top graph: Light intensity (in µmol photons m^{-2} s^{-1}) above the canopy. Bottom graph: light attenuation in an oak forest in early February (left hand side) and end of July (right hand side). Please note different scales for color bars. All units are in µmol photons m^{-2} s^{-1}.

Wind and Temperature

Wind redistributes heat and mass, such as water vapor, carbon dioxide, and other gases in the atmosphere. Generally, in forest canopies, just like light, wind is attenuated due to the resistance of the stems, branches and leaves. More specifically, wind speed and leaf size, shape and distribution can influence the exchange of CO_2 and water at the leaf level (small scale) and at the canopy level (large scale) by affecting boundary layer conductance of leaves or canopies. At the leaf level, the boundary layer is a very thin layer of stagnant air surrounding the surface of a leaf. The thinner it is the less resistance it offers to the movement of water and CO_2 molecules. One way in which plants can reduce their leaf boundary layer is having narrow leaves, such as pine needles, or leaves with large indentations, such as pin oak leaves. Wind increases leaf boundary layer conductance by reducing the thickness (and therefore the resistance) of the boundary layer as wind speed increases.

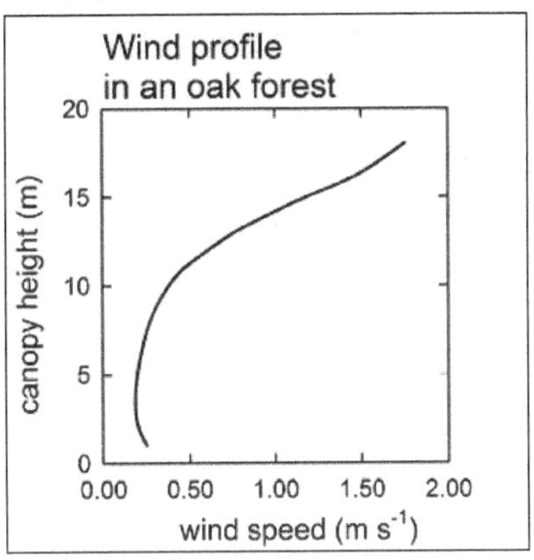

Wind profile through an oak forest in the New Jersey Pine Barrens throughout a year.

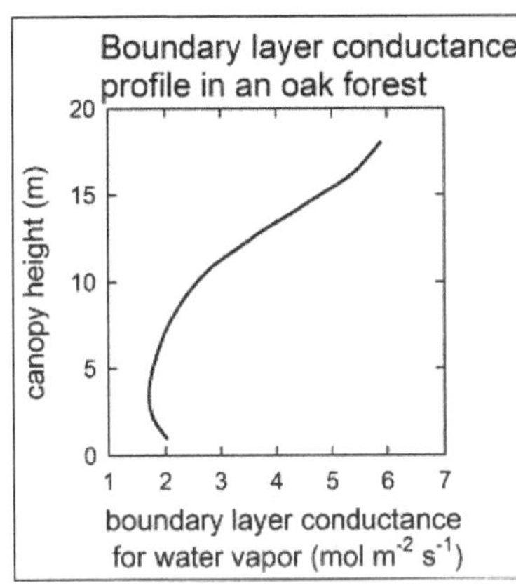

Boundary layer conductance profile through an oak forest canopy in the New Jersey Pine Barrens.

The boundary layer of the canopy is determined by the shape, density, and aerial spread of the crowns of trees in a forest. Together these qualities of the canopy are the roughness. Widely spread conifers with pointy tops have a higher roughness than oaks with round tops and denser packed leaves. The rougher the canopy, the greater the wind turbulence it creates, and the more wind (and therefore CO_2 and water vapor) moves between the leaves of the trees. Thus, conifers tend to have thinner leaf and canopy boundary layers, and oaks tend to have thicker leaf and canopy boundary layers. When boundary layers are thin, leaves and the air surrounding them are more coupled (i.e., the wind has a larger impact on leaf physiology), and when the boundary layers are thicker, the leaves and the surrounding air are decoupled (less direct impact on leaf physiology).

The impact of the boundary layer on leaf physiology is linked to the vapor pressure gradient, which is the difference in water vapor concentration between the inside of the leaf and the air surrounding the leaf. A thicker boundary layer reduces the vapor pressure gradient by increasing the resistance to water vapor movement. Less water vapor movement away from the leaf means less evaporative cooling, and the leaves get warmer. If leaves get too warm, the process of photosynthesis breaks down, and the tree suffers. Thus, oak leaves at the bottom of the canopy can be wider (which increases their boundary layer thickness) and horizontally oriented (which increases their light absorption) even though there is less wind, because there is less light at the bottom of the canopy and therefore less heating of the leaves. At the top, the leaves need to be narrower and more vertically oriented to reduce the boundary layer and heat absorption. In conifers, because of the way the leaves and crowns are constructed, all the leaves have thin boundary layers. The thinner the boundary layer, the more important stomatal control of water loss becomes, because water vapor is more likely to move out of the leaf (greater vapor pressure gradient).

Jarvis & McNaughton have quantified the influence of the atmosphere on gas exchange in forest canopies with a decoupling factor Ω. It describes the linkage of the physical environment with the biological processes in plant canopies and is defined as:

$$\Omega = \frac{\varepsilon + 1}{\varepsilon + 1 + \frac{g_b}{g_s}}$$

The decoupling coefficient Ω is unitless, and falls between 0 and 1. The rougher the canopy, the greater boundary conductance (gb) becomes, the closer Ω is to 0, the more coupled the canopy is to the atmosphere, and stomatal control becomes more important. In contrast, the less rough the canopy, the smaller the boundary layer conductance becomes, Ω is closer to 1, the canopy is decoupled from the atmosphere, and stomata conductance has less impact on transpiration.

FOREST RESOURCES

Forests provide an array of benefits to human societies above and beyond their pivotal roles as habitat and environmental regulators in natural ecosystems. These benefits are often described as resources that people can draw upon for fuel, lumber, and recreational or commercial purposes. The perception that forests provide resources for people has been a prominent factor in spurring efforts to preserve forests. The signs that welcome traveling tourists at the entrances of the White Mountain National Forest in New Hampshire read "Entering the White Mountain National Forest: land of many uses." When the U.S. government set up the National Forest system, it was eager to communicate that the land was not simply being taken out of commercial development; rather, there were many practical and productive reasons for preserving these woods.

A growing awareness on the part of governments and the general public, in the United States and around the world, of the benefits of forests to humans has given rise to government agencies and a thriving industry devoted to forest resource management. The mission of forest resource management is to develop, protect, and manage the multiple resources of forests through professional stewardship, enhancing the quality of life for the public while ensuring the conservation and sustainability of these resources.

Issues and Impacts

Forests provide clean water and air, timber for wood products, wildlife habitats, stable soil, and recreational opportunities, and they beautify the environment. Furthermore, they are also an important economic resource producing marketable timber. In spite of this economic value, much forestland is being taken out of forest production, largely due to urban sprawl, which has claimed forested areas that will never be replaced. On the other hand, lands that have been designated as wilderness areas or parks will continue to provide many environmental and scenic benefits, but will not produce wood products. In the United States, governments and industry concerned with forest resource management are particularly concerned about the thousands of acres of underproductive forestland that is cut over for quick cash each year, and then left idle. Thus, poor forest management can lead to long-term economic losses. The unrealized economic potential of forests has prompted a concerted effort at national and local levels to ensure that forests be well managed for environmental, aesthetic, or economic reasons rather than left unproductive and inactive.

ROLE OF FORESTS

The Earth's forest cover provides a natural protection to all of us. Forests are home to numerous wildlife species. At the same time, they help us meet our social and economic needs. Forests also help regulate the environment. The main functions of forests are protective, regulative and productive.

- Forests act as a natural habitat for a variety of flora and fauna. Without the forest cover, a number of species would have no place to survive. Thus, by supporting so many species, forests also protect biodiversity. Due to deforestation activities, a large number of species of flora and fauna are endangered.
- Forests help maintain the climatic environment of the Earth. They regulate climatic temperature.
- The trees in these forests absorb the carbon dioxide that is released into the atmosphere. Since forests remove carbon dioxide, they help to bring down the pollution level in our surroundings and reduce global warming.

- Trees in the forests give out oxygen in the environment. Oxygen is vital for life on earth to continue.
- Forests act as a natural cover for the soil's surface. These natural catchments do not allow the rainwater to take away surface soil. Thus, they prevent soil erosion on one hand and siltation in rivers and seas on the other hand.
- Forests also control floods and their hydrological cycle avoids calamities like droughts.
- A number of raw materials required by industries such as the paper and mining industries are found in the forests.
- Forests provide for economic needs of human beings, such as fuel and food supplies.
- Forests act as a source of livelihood for a number of people who depend on them for fruits, honey, medicinal plants, sometimes even shelter.

Productive Role

It is estimated that the forest products contribute about 1% of world gross domestic product (GDP). The annual turnover of timber and other wood products from forests is valued at more than US$200 billion. The demand for commercial timber and other products is ever increasing, and expected to rise by 50% by 2010. Apart from that, nontimber products like rubber, cotton, medicinal products, food and so on represent significant economic value.

Food

Rhizome	Amarphous campanulatum, Cyprus rotandus
Root and aerial	Dioscorea , Moringa oleifera Caryota urens, Bauhinia variegata.
Buds	Dillenia pentagyna, Phoenix sp.
Sap, and latex, Bark	Borassus flabellifer, Cissus rapanda.
Stems	Cycas pectinata, Dendrocalamus strictus.
Leaves	Tamarindus indica, Moringa oleifera.
Flowers	Ficus glomerata, Madhuca indica, Bambax ceiba, Tamarindus indica.
Fruits	Aegle marmelos, Anacardium occidentale, Anona squamosa. Artocarpus heterophyllus,Borassus flabellifer, Capparis decidua, Diospyros melanoxylon, Emblica officinalis, Ferronia elephantum, Morus alba, Zizyphus spp.
Seeds	Anacardium occidentale , Juglans regia, Prunus amygdalus, Tamarindus indica, Dendrocalamus strictus.

Fuel

Wood is used as fuel for thousands of years, until the advent of coal, oil, gas, electricity, etc. Wood constitutes as chief source of fuel. Even today more than half of the total world consumption of wood is for fuel-wood. Wood remains the major source of domestic fuel. Approximately 175 mm3 of wood is used as fuel in the country. It is estimated that by 2010, most of the 3 billion people who depend on it for their daily living will find it hard to obtain. Already, rural families spend precious hours in collecting firewood instead of other productive work, something that causes losses to the tune of US$ 50 billion to the world economy.

Eg., Acacia spp, Casuarina equisetifolia, Prosopis, Neem, Leucaena leucocephala, etc.,

Shelter

Wood is used for construction of buildings.

Eg., Palmyra, Teak, Jack, etc.,

Clothing

Rayon cloth eg., Eucalyptus spp.

Timber

Timber is a major forest produce and is used extensively for various purposes. Most of the wood produced is used for construction of houses, agricultural implements, bridges, sleepers etc., 12 mm3 of timber is produced from our forests. More than 1500 species of trees are commercially exploited for timber in different parts of India. It is used in timber-based industries such as plywood, saw milling, paper and pulp, and particle boards.

Many species like teak, sal, deodar, babul, sissoo, chirpine, adina, axlewood, rosewood, dipterocarpus, and etc.yield valuable timber.

Industrial Wood

Forest provide raw material to large number of industries eg: paper and pulp, plywood and other boards, packing cases, matches, toys etc.,

Paper and pulp	Bamboos, Eucalyptus, casuarina
Plywood	Teak, Rose wood, Terminalia etc.
Packing cases	Pinus sp, Silver oak, Fir,
Matchwood	Ailanthus, Simaruba, Bombax
Toys	Adina, Redsanders, rose wood

NTFP

- Fibre and flosses: Fibres are obtained from bast tissue of certain woody plants, which are used for making ropes. Flosses are obtained from Ceiba pentandra and fibres are obtained from Agave sisalana, Sterculia urens.
- Grasses and bamboos: A large variety of grasses are found in the forests. About 30% of the 416 million livestock population graze in the forests. Among valuable grasses eg: Sabai (Eulaliopsis binata) is harvested annually 6.5 million tones and 80,000 tonnes of bamboo are harvested from forest every year.
- Essential oil: India produced about 1500 tonnes of essential oils during 1980, which was utilized in making soaps, detergents and chemicals eg. Eucalyptus, Bursera, Cymbopogan, Santalum album etc.
- Oil seeds: Many tree species of Madhuca indica, Pongamia pinnata, Shorea robusta, Azadirachta indica, Schleichera oleosa, Vateria indica etc., produce oil-bearing seeds, which are commercially important. Presently these seeds are used in the soap industry. There is a potential production of about 1 million tonnes of oil every year from forests tree seeds.
- Tans and dyes: Important tannins are extracted from myrobolan nuts, bark of wattles (A.mearnsii, A.decurrens, A.dealbata) and Cassia auriculata , leaves of Embelica officinalis and Anogeissus latifolia, bark of Cleistanthus collinus, fruits of Zizphus xylophora, Cassia fistula, Terminalia alata, T.arjuna etc., katha and cutch are obtained from Acacia catechu.
- Gums and resins: Gums and resins are exuded by trees as a result of injury to the bark or wood. Gums –eg: Sterculia urens, Anogeissus latifolia, Lannea coromandalica, Acacia nilotica, Ptercapus marsupium, Butea monosperma etc.,

Resin is Obtained from Pinus Roxburghii

Spices	Seeds of Carum carvi, barks of Cinnomomum zeylanicum, dried capsules of Elletaria cardomomum.
Insecticides	Pyrethrum and neem

- Drugs, Spices and Insecticides: Important spices yielding drugs are Rauvolfia serpentina, Hemidesmus indicus, Dioscorea spp, Atropa spp, Datura innoxia etc.
- Tendu and other leaves: Tendu leaves (bidi leaves) (Diopyrus melanoxylon) and leaves of bauhinia spp, Butea spp, plates, dona etc.
- Lac and other products: Lac is a resinous secretion of insects which feed on forest trees eg ; Butea monosperma. Silkworm is feed on Morus alba or Terminalia alata. Honey is produced from forests.

- Fodder and grazing: About 30% of 416 million livestock population depend up on forest grazing and leaf fodder supply.Eg; Luecaena leucocephala, Albizzia lebbeck, Hardwickia binate.
- Cane: Canes or rattans are the stems of a climber plant and are used for a large number of household items. It is used to make walking sticks, polo sticks, baskets, picture frames, screens, and mats.

Environmental Protection

Forests play an important role in environmental protection. There is a long history of protection forests in mountain areas, where they help to prevent soil erosion, landslides and avalanches, and where they are important in maintaining the water quality of rivers draining forested catchments. Special silvicultural methods are required to ensure that these forests are maintained indefinitely. Forests also respond to environmental protection. A major issue is air pollution, which is known to have had significant impacts on some forests. Air pollutants of concern include sulfur dioxide, hydrogen fluoride, heavy metals, and ozone. Control of these pollutants ultimately benefits forests. Forests have a major role to play in the protection of the global carbon cycle. They represent an important sink for atmospheric carbon dioxide, and conversion of forests to other land uses is one of the causes of the increase in atmospheric carbon dioxide concentrations. Reforestation and afforestation could contribute to reducing atmospheric carbon dioxide concentrations, and the use of biofuels could help to reduce demand for fossil fules.

The relationship between forests and the environment has been recognized for more than a thousand years, yet forestry practices continue to cause damage to the environment in the form of soil erosion, water quality deterioration, and other adverse effects. Some of the earliest records of problems associated with the removal of forests come from Japan, where logging of the montane Cryptomeria japonica forests more than 1000 years ago was accompanied by increases in the incidence of flooding in lowlying areas. Since then, forests have continued to be cut in headwater areas, often with disastrous consequences downstream. The value of forests as a means of environmental protection has, however, slowly been acknowledged. In areas where mass movements are common, and there is a relatively high population density, such as the European Alps and Japan, forests have been formally designated as having protection as their primary role. Elsewhere, the role of forest cover in protecting water sources and other values has also been recognized. Links between forests and the atmosphere have been identified. As the control of pollutants and other substances altering the composition of the atmosphere becomes increasingly important, there has been a growing acknowledgement of the role that forests can play in global environmental protection. Forests are both affected by the pollutants and can themselves play a role in altering the atmospheric composition. Consequently, environmental protection measures taken to protect human health may have beneficial effects on forests. Large-scale afforestation as a means of reducing atmospheric carbon dioxide concentrations has been discussed, and some countries (e.g., Australia) are already encouraging such a policy. In this article, various aspects of the relationship between forests and environmental protection are examined. Examples are given from the boreal, temperate and tropical zones, in order to emphasize that environmental protection is an issue facing forestry in all parts of the world.

Protection Forests

Protection forests have as their prime function the protection of the environment. They are common in mountainous areas—where they stabilize slopes, prevent avalanches, and protect water quality, and also in coastal areas, where they stabilize sand dunes. For example, in Austria, 741 452 ha of forestland have been classified as protection forest, representing 19.2% of the total forest area; these protection forests typically occur on steep (average gradient ca. 65%) slopes, and often occur at above 1800 m in altitude. Elsewhere, the information on the amount of protection forest is very fragmented because the definition of protection forest varies between countries. For example, in Greece, 100% of the forests are classified as being managed primarily for soil protection, yet this does not preclude the annual removal of >2 million m3 (overbark volume) of wood from the forest estate. In many tropical countries, certain forests have been formally designated as protection forests, but this designation is not always adhered to on the ground.An example of the role of protection forests in maintaining water quality is provided by studies in the Oxapampa municipal watershed, near Cerro de Pasco, Peru. The area lies in an area characterized by tropical montane forest, and the landscape consists of steep slopes that are susceptible to erosion. Annual soil losses were compared between areas covered by natural forest, areas with a mixture of crops, pasture and forest, and areas of pasture only. The maximum suspended sediment concentrations reached 76, 226, and 771 mg/L, respectively, in the three land-use types during storm flows. Average annual soil losses from the three land-use types were estimated to be 121, 345 and 542 kg/ha, respectively.

Protection forests present particular problems for management. In Switzerland, in those protection forests where timber extraction is permitted, management has been described as "near-natural," involving either single-tree selection and continuous cover techniques or very small (<0.1) patch cuts. In the French Alps, the ideal structure of protection forests dominated by Picea abies is believed to be uneven-aged stands with gaps. Management activities often aim to increase stand heterogeneity, and uneven-aged stand management has been adopted in some North American protection forests, such as the Quabbin Forest of central Massachusetts, which is an important source of Boston's water supply. However, in some parts of the world, commercial felling within protection forests remains an important economic activity, and management is concentrated more on obtaining the best possible growth, while maintaining a forest cover. Thus, in the Russian Far East, forestry operations take place in the mixed, uneven-aged, protection forests. The fellings are designed to reduce the proportion of Quercus mongolica, which is usually the dominant species in the overstory, and other broadleaved species. The management objective is to increase the proportion of Pinus koraiensis, a species that has a higher volume increment than Quercus mongolica. Such major interventions are not typical of European protection forests, but may be relatively common in areas where protection forests are still seen as an important source of timber.

Often, the frequency and extent of silvicultural intervention is strictly limited by law, a regulation that is based on the assumption that undisturbed forest ecosystems are stable. This is now known to be untrue: forest ecosystems are highly dynamic and subject to regular disturbance. As a disturbed protection forest may no longer fulfill its role, management is usually required to reduce the potential impacts of disturbances. Instances where protection forests lose their protective roles include those damaged by forest fires and by large-scale insect outbreaks, such as occurred in Switzerland as a result of major storms in 1990 and 1999. The 1990 storm caused major problems for foresters throughout Switzerland, and in other alpine countries. The amount of fresh dead wood was such that complete removal was impossible.

This resulted in the buildup of bark beetle populations, further aggravating the instability of the forests. The problems may also have been exacerbated by the laissez-faire approach to management that has resulted in the stands containing a high proportion of very old trees and inadequate regeneration. In some cases, the associated high levels of dieback and mortality were attributed to air pollution, a convenient explanation that shifts the responsibility for forest instability away from the forest manager. In another example (in the Swiss valley known as the Valais), the Pinus sylvestris forests situated on the slopes above the town of Visp have been severely impacted by Tomicus spp. This insect species normally does not directly cause the death of trees, as populations are generally insufficient to affect trees severely. The cause of the severity of the pest outbreak in the Valais is uncertain, but may involve a combination of severe climatic stress, damage caused by fumes from a nearby industrial plant, very heavy mistletoe (Viscum album) infections, and the creation of good insect breeding habitat in the form of small groups of trees blown over by a severe windstorm. The effect is that the beetles have spread, and many trees have died. Salvage logging has been undertaken by conventional road-based logging, and also by the much more expensive helicopter logging. The challenge now is to ensure that the gaps created by the salvage logging do not result in snow and debris avalanches that might endanger people living in the town below the forest.

Assessments of the ecological stability of protection forests are still undertaken, and form an important step in the development of management prescriptions for such forests. A recent case study of the Ban de Ville forest in Courmayeur (Aosta Valley, Italy), revealed many problems typical of alpine protection forests today. The 143 ha forest faces west, and is situated at altitudes between 1300 and 2300 m. It is dominated by Picea abies, although Larix decidua becomes increasingly common above 2100 m. Two thirds of the forest was found to be "unstable," with identified problems including unsuitable stand composition, an oversimplified vertical structure and cover, the presence of high densities of wild ungulates, and the presence of forest pests (mainly the bark beetle Ips typographus) and diseases (mainly Heterobasidion annosum). Steps to reduce these problems included increasing the proportion of Larix decidua in the pure Picea abies stands and gradually establishing a multi-layered, small group structure similar to that proposed for Picea abies protection forests in the French Alps. Reestablishment of protection forests following disturbance can present major problems. The forests grow close to the treeline, and are therefore, close to their ecological limit. The microclimate in a forest is very different to that outside the forest, being more favorable to regeneration. Once the canopy cover is gone, the climatic conditions are much harsher, and it may be very difficult to establish a new tree cover. Consequently, it is almost impossible to reestablish the same species composition in a heavily disturbed protection forest. An example of such a system is the protection forests in the Urseren valley, near Andermatt, Switzerland. Here, the forests currently consist of Picea abies and Larix decidua, with some Alnus viridis scrub. While regeneration is possible within the forest, regeneration would be impossible in the event that the forest cover was destroyed.

A particular problem is sliding snow. In a mature forest, this process is limited by the presence of trees. When these are gone however, the buttressing effect is largely removed and snow avalanches can more easily occur. Additionally, stems of young trees may be severed during avalanching. There are various strategies that can be adopted to reduce the problem. If the trees in a protection forest have died, they can be cut at a height of 50 cm to 1 m above the ground surface, leaving high stumps, and thus some buttressing potential. Alternatively, artificial barriers can be installed, although these may be costly. During the initial planting phase, a variety of techniques may be used to reduce snowslide, including stakes, tripods and horizontal logs 18–20 cm in diameter. In Europe, these barriers can be metal and concrete, but frequently, they are constructed from wood. The wood from the surrounding forest can be used, but more often than not, such wood has a low mean life. For example, the heartwood of Larix decidua and Pinus sylvestris has a mean life of 8–15 years, which may be insufficient time for sufficient regeneration. The heartwood of Castanea sativa and Robinia pseudoacacia is more resistant to decay, but usually has to be brought in from elsewhere. It is also possible to ameliorate the impacts of sliding snow by planting trees that are less susceptible to this form of damage.

In extreme cases, very few trees are capable of growing, although Pinus mugo and Alnus viridis have been used successfully in the state of Bavaria in Germany. Clumps of Larix decidua, Fagus sylvatica, and Acer pseudoplatanus have also been used. Sorbus aria and Sorbus aucuparia are also fairly resistant to damage from snowsliding. These stabilize the snowpack, enabling the more sensitive Picea abies and Abies alba to be planted. Throughout British Columbia and southeast Alaska, red alder (Alnus rubra) vigorously colonizes snow avalanche tracks and is resistant to moderate-sized avalanches. Another problem is the presence of very high populations of game (red deer, roe deer, chamois, and ibex) in some protection forests. These artificially high populations have arisen because of hunting policies that allow very high populations to exist (often involving feeding animals in winter and encouraging more palatable forage). In addition, recreation activities above the treeline are increasingly disturbing animals, so that they spend longer in the forest. The result is very high grazing pressure; in some areas this can severely affect forest regeneration. This problem seems to occur throughout the alpine region, but is particularly apparent in Bavaria (Germany), Austria and Switzerland. It is actually a problem in many other types of European forest, and is not restricted to the Alps.

Protection forests do not only occur in mountainous areas. For example, in China, they are used extensively for protection against erosion in semiarid hill-gully systems. An example is provided by the "Three Norths" protection forest system in the northern part of the country. Plans for this system were passed by the State Council in 1978, with the first stage of development being completed in 1985. In establishing this system, account was taken of the ecological and economic benefits of the scheme, and the importance of rehabilitating and protecting the environment was given greater importance than timber production. In many areas covered by the scheme, forest cover has been increased dramatically. For example, forest cover in one demonstration catchment on the loess plateau in Shanxi has been increased from 9.17% in 1977 to 24.4% in 1988. This is all the more remarkable given that the 8917 km2 catchment has a population of 450 700 people. In another area, soil loss has been reduced by more than 95%, and runoff has been reduced by 91%. These changes have been accompanied by better agricultural planning, resulting in greater productivity of farmland, greater productivity of animal husbandry, and a 250% increase in per capita income. Integrated development projects of this nature, where protection forests are one component of a series of improvements, are clearly of considerable potential. However, recent large-scale "soil and water conservation" projects on the loess plateau supported by World Bank funding are promoting the production of high cash value agricultural products (both crops and orchards) on large constructed terraces at the apparent expense of the long-term sustainability of semiarid soil resources.

Another area where protection forests are of value is in coastal situations. Protection forests are used to stabilize sand dunes, but they are also used to protect muddy shorelines. In tropical and semitropical areas, mangroves are important as coastal protection forests. Very often, only a very small strip of forest remains. However, to fulfill protective functions, a mangrove forest needs to be quite wide. For example, in Indonesia, a greenbelt 614 m wide between the sea and forest fishponds has been suggested as necessary. Coastal protection forests have been used for a long time. For example, the Etang-Sale forest was established on the island of Réunion in the 1870s in an attempt to stabilize the coastal dunes. The forest was planted with Casuarina equisetifolia, although more recently, a variety of other species has been used, including Acacia auriculiformis, Azadirachta indica, Eucalyptus camaldulensis and Khaya senegalensis, with Albizia lebbeck planted as an understory species to increase the vertical diversity. Coastal protection forests differ from mountain protection forests in that they are often planted on land that has never previously been forested. Consequently, exotic species are more often used than in mountain situations. For example, Pinus nigra (a pine species originating in southern Europe) has often been planted on sand dunes in Britain.

Many protection forests are designed to reduce the incidence of mass movements, particularly snow avalanches. However, forests may also help increase slope stability. This is because the roots of trees play an important part in increasing the strength of soils on slopes. If forests are removed, the roots will decay quite rapidly, with root strength normally at a minimum three to 15 years following clear cutting. The reduction in strength may be accompanied by landslides and other forms of slope instability. In addition to landslides, the establishment of forests can also result in splash erosion. If there is very little vegetation cover under the forest, drip from the canopy may be sufficiently large and have sufficient energy to cause erosion of the soil. This is particularly true in some tropical plantations of Tectona grandis, especially where the leaf litter is removed. In such situations, it is possible to reduce the erosion by maintaining a ground cover, and by allowing the development of a thin organic horizon. Despite these particular problems, there is a great deal of evidence that a well-managed forest will reduce the amount of surficial erosion, especially if it contains multiple strata (which help reduce the velocity of falling drips). The key to managing erosion in plantations is to maintain the infiltration capacity of the soils during the life cycle of the trees.

Influence of Forests on Climatic Conditions

The forest environment or site consists of the physical environment surrounding the aerial portions of the tree (climatic factors) and that surrounding the subterranean portion (edaphic factor) and the third one is biotic factor. External influences, particularly fire, grazing and browzing animals and humans, affect markedly the nature of sites and their capacity to support tree growth.

As the forest becomes established and develops, the site itself is greatly changed. Forest cover moderates the extreme diurnal temperature regime of open sites resulting in more uniform conditions. Wind velocity is slowed in the vicinity of tree crowns and becomes negligible within the forest.

Trees crowns intercept sunlight and alter the quantity and quality of radiation reaching forest floor compared with that reaching open sites. On the forest floor, accumulating layers of leaves, twigs and other litter attract a characteristic grouping of plants and animals that live on decaying organic matter and on each other.

Biotic Environment

Ablotic Environment

The physical environment controls the organisms, and organisms also influences and control the abiotic environment in many ways.

Influence on Air Temperature

Forest vegetation reduces mean annual temperature five feet, above ground from 0.8 to 1.8FO depending upon character of forest and locality, particularly elevation.

The effect of a forest in reducing temperature Is much greater at 5 feet above ground than in the tree crowns.

The following table shows the extent to which species influences the effect of vegetation on air temperature.

Concurrent conclusion of various investigation on the effect of forests on air temperature are:

- Forest lower the daily mean temperature in spring and summer and raise it slightly in autumn and winter.
- Forest lower the daily maximum of air temperature and raise the daily minimum.
- Forest diminish the daily range of air temp. 2.2. Influence on Precipitation.

Air of forest is cooler and moister, than a-r in the open. Forest increases the precipitation of any 3rea. As air cools in rising, precipitation increases with increase in elevation. The effect of forest vegetation on precipitation relates to its effect on:

- Local precipitation.

Influence on Local Precipitation

The influence of forests on local precipitation at low elevation is negligible but their influence increases rapidly with increase in elevation particularly in mountainous regions.

The forested mountains increase precipitation with increase in elevation to a much greater degree than denuded mountains. Effect of forest on precipitation varies with the species, coniferous trees having a greater influence than broadleaved species.

Influence on General Precipitation

The influence of forest on general precipitation is far greater than their local effect. Forest cover profoundly influences evaporation of water from the land. 2.3. Influence on Air Currents.

A forest break or moderate the force of air currents they serve to protect lands lying to their lee against cold and dry winds and against winds of high velocity. three parallel lines of trees planted at 3x2m intervals.

Influence on Atmospheric Humidity

The effect of forest vegetation .n the relative humidity of the air appears to vary considerably in different localities, more particularly with differences in elevation. 2.5. Influence on Evaporation.

Vegetation and particularly forests, in checking the velocity of wind at and near the soil surface, in slightly raising relative humidity of the air in shading the soil, In lowering temperature and in covering the mineral soil with a mantle of leaves and other litter, retards rapid loss of moisture from mineral soil.

Loss of moisture through evaporation from surface soil in the open on a windy day may be five times as great as loss of moisture from forest soil of similar character under protection of forest cover. Evaporation from snow surface in the open may be four times as rapid as from similar surface protected by a forest cover.

Influence on Transpiration Loss

A large part of the water absorbed by vegetation is taken from the soil, enters the transpiration current and is returned to the air through the leaves. Forests often lower the water level in the soil and that The layer of Forest soil below the surface are usually dryer during the growing season.

Influence of Forest on Edaphic Conditions

Influence on Soil

The beneficial influence of forest vegetation on soil is due to its beneficial effect and its power of increasing fertility of the land by adding nutrients.

In general the influence of forest vegetation on soil related to -the producing of a new substratum of soil and the changing of soil structure. Forest vegetation assists in the formation of soil by the accumulation of plant remains by stimulating weathering through the action of acids formed by vegetation, and by the resistance which forest vegetation offers to moving air and water.

Influence on Soil and Temperature

Forest usually reduce the maximum soil temperature and increase the minimum soil temperature with the depth of the soil.

The influence of the forest in reducing the maximum soil temperature results partly from the shade of the crowns and partly from the insulation of the forest floor. The influences of forest vegetation on freezing of soil are of great importance. Soil under a forest usually remains soft when that in the open is frozen to considerable depth.

The distance beyond the edge of the forest's influence on the soil temperature is about the same as the distance of the influence on radiation.

Influence on Seepage and Water Retention

Forest vegetation, by reducing surface runoff, increase the amount of water that percolates into a soil. The effect of forest in preventing freezing of forest soil of vast importance in increasing the amount of water that percolates into it, particularly during the spring months.

Forest cover in increasing the volume of soil in mountainous region over the solid rock foundation, increase seepage. The humus layers, characteristic of every well-managed forest, absorb from two to four times their weight of water. Forest soil, with its overlaying organic layers is in a real sense a vast sponge capable of absorbing much more water per unit area than soil in the open. Therefore water-holding capacity of humus rich soils is highly increased.

Influence on Wind Erosion

Soil is transported from one place to another either by wind or by water. Forests, by checking the velocity of wind and by reducing surface runoffhave a great influence on the stability of soil. Sands subject to wind erosion should be covered with forest growth or their soil binding plants because permanent, stability can be attained only where sands liable to shift are so covered.

Influence on Water Erosion

One of the most important and far-reaching influences of the forest is the protecting of the soil from washing. Forest plays an important role in erosion control owing to the combined effect of:

Influence on Springs

Part of the water that seeps into soil at higher elevations reappears at the surface lower down in springs. A forest through its influence in increasing seepage and decreasing, surfaces runoff provides a large supply of ground water, particularly in mountainous and hilly region, for the feeding of springs. In level contrives where the general effect of an forest is to drain the soil and lower the ground water, springs seldom cover and the effect of the forest upon them is of minor importance.

Influence on Floods

Forests, in reducing surface runoff and increasing seepage, extend the time over which precipitation reaches as streams. The most striking influence of forest vegetation on stream flow is shown where heavy.

Influence of Forest on Biotic Conditions

Influence on Animal Life

Plants are probably the most important feature in the environment of terrestrial animals. All animals are dependent directly or indirectly upon plants for food. Forests also provides many animals with shelter from inclement weather, protection from enemies, situation for homes, and materials for nests.

Forests as Source of Food for Animals

Many forest animals feed very largely upon seeds, fruits, buds, leaves, twigs and other parts of forest plants, other feed in forests when food is plentiful there and scarce elsewhere.

Forest as a Source of Shelter and Protection

Forests provide protection for most terrestrial 'game birds. classed as game are more or less dependent upon forest for food and protection.

Influence on Mankind

Influence on Civilization

Forest have had an important influence on the distribution of mankind over the earth's surface. Three stages in the relation of human progress to forests.

- Civilization dominated by forests.
- Civilization overcoming by forests.
- Civilization dominating by forests.

In regions where forests are abundant, logs and lumber are usually so dwelling; but in regions leaving forest or where forests products are scare.

Favorable Influence on Weather

Carbon Sink

Forest have the potential to contribute to climate change through their influence on the global carbon cycle. Forest are being recognized as playing important roles in global biochemical cycle. Major pools of carbon are the atmosphere, fossils fuels, oceans and terrestrial biota and soils. Forest have the potential to influence global warming in other ways too. Carbon pools in forest.

The world's forests contain more than 55% of the global carbon stored in vegetation and more than 45% of that in soil. Most of the carbon Pools in forest vegetation is located in tropical forests (62%), whereas most of the carbon pool in forest soils is located in boreal forests (54%).

To prevent environment it is essentially to prevent carbon emissions by conserving existing carbon pools in forests and establishing plantation on non forested lands.The adoption of forest management options that conserve and sequester carbon would help to prevent forests from becoming a significant net sourceof Co2 for the atmosphere in the future and thereby help to offset other factors that contribute to accelerate global warming.'

Experiments have repeatedly d3monstrated an increased productivity of crops and trees when irrigated with waste water as compared with clear water. The application of waste water at the rate which ensure a balance between nutrient input and plant uptake will promote optimal plant growth while limiting the risks of pollution.

Influence of Forest on Pollution

Trees and other green plants produce oxygen and consume carbon dioxide. Major pollution such as water, air and noise pollution can be controlled through forests.

Air Pollution

According to who, air pollution may be defined as:

A wide range of inorganic gases, organic compounds, inorganic metallic substances and root particles, is discharged into the atmosphere by motor vehicle, factories, power plants, have furnaces and waste incineration plants.

Following are the major air pollutants:

- Gases.
- Carbon Monoxide. CO.
- Inorganic Metallic Substance.

Control of Air Pollution through Forests

Plants materials are among the most effective air conditioners in existence. Under present circumstances and grave situation of air pollution it appears that plant as a great source of natural air conditioning is being overlooked.

Forests play an important role in climate amelioration by reducing the intensity of solar radiation, modifying temperature, reducing evaporation from soils and increasing the amount precipitation.

Plants as Air Cleaner

Plants Perform an important role in reducing air pollution and thus they act as cleaner.

- Narcosing or wind turbulance.
- Masking.

Water Pollution

Causes of Water Pollution

Water pollution arises from a variety of sources industry, domestic sewage, agricultural fertilizers and feedlots, construction activities, and forest practices. Toxic chemicals dumped by both industrial units and government agencies are the major source of water pollution. These chemicals are greatest threat to fresh water. - Pesticides

Control of Water Pollution through Forests

Plants reduce pollution in water and can hold top soil in place, thereby reducing the sediments and nutrients as a pollutant to water resources in the following ways:

Erosion Control and Watershed Protection

Plants use in conservation practices reduce water cause soil erosion by intercepting rainfall, holding soil with their roots, and increasing water absorption through addition of organic matter. 'thus plants are more attractive than mechanical water erosion control devices.

In forest, pesticide or insecticides are commonly used to control harmful insects and pests. Salts occur naturally in soils and waters. In humid regions, rain readily leaches salts out of the soil. The salts present is the water are thus the main source of salt accumulation in those soils. Plants. Many fruit trees also absorb salts through their leaves.

Noise Pollution

Experts refers to noise as “Invisible pollution”.

Causes of Noise Pollution

Today traffic is the major source of human communication, which ultimately lead to noise pollution.

Control of Noise Pollution through Forests

- Wind Belt: Plants mask unwanted sounds.
- Proper Highway Design: Tree species did not differ greatly in their ability to reduce noise levels, provided that deciduous trees are in full leaf.

- Well design planting in such situation can aid tremendously in sound control. It is concluded that plants play a vital role in noise abetment. They can become more effective in playing their role of noise control if planted in massive form. Thus trees, shrubs and all other vegetation aid in noise pollution reduction if used properly.

FORESTRY

Forestry is defined as the theory and practice of all that constitutes the creation, conservation and scientific management of forests and the utilization of their resources. It includes all thinking and all actions pertaining to creation and management of forests, including harvesting, marketing and utilization of all forest products and services. It includes not only management of existing forests but also the creation of new forests.

- Protection Forests: The practices of managing the forests for their protection function are called Protection forestry. In protection forestry, the object is to protect the site due to instability of terrain, nature of soil, geological formations, etc. Such areas where manipulation of the forest cover is not desirable may be classed as protection forests. The forests located on higher hill slopes, national parks and sanctuaries, preservation plots, biosphere or nature reserves and wilderness areas may be included under protection forests. The practice of forestry with a view to conserving flora, fauna, soil and water, increasing water yields, reducing floods and droughts, amelioration of climatic conditions, etc. is called Protection Forestry.
- Production Forests: The practice of forestry with object of producing maximum quantity of timber, fuel wood and other forest produce is called Production Forestry. The production forestry can be further classified into:
    - Commercial Forestry: Commercial forestry aims to get maximum production of timber, fuel wood and other forest products as a business enterprise.
    - Industrial Forestry: Industrial forestry aims at producing raw material required for industry. In Production forestry, there is a greater concern for the production and economic returns.
    - Mixed Quality Forests.
    - Valuable Forests.
    - Inaccessible Forests.
- Social Forests: Social Forestry is the practice of forestry which aims at meeting the requirement of rural and urban population. The object of social forestry is to

meet the basic needs of community aiming at bettering the conditions of living through:

- Meeting the fuel wood, fodder and small timber requirements.
- Protection of agricultural fields against wind.
- Meeting recreational needs.
- Maximizing production and increasing farm returns.

Various forms of Social forestry are:

- Community Forestry: The practice of forestry on lands outside the conventional forest area for the benefit of local population has been called Community forestry. Community forestry seeks the involvement of community in the creation and management of such forests.
- Farm Forestry: Farm Forestry is defined as the practice of forestry in all its aspect on farms of village lands generally integrated with other farm operations.
- Extension Forestry: Extension forestry which includes the activity of raising trees on farm lands, villages wastelands and community forest areas and on lands along the sides of roads, canal banks and railway lines. More recently, there has been emphasis on dynamic land use planning and efforts are made to maximise production on farmlands under agroforestry.
- Agro-Forestry: Agro-forestry has been defined as a sustainable land management system which increases the yield of the land, combines the production of crops and forest plants and animals simultaneously or sequentially on the same unit of land and applies management practices that are compatible with the cultural practices of the local population.
- Recreational Forestry: More recently, there has been considerable demand for Recreational forestry, which is defined as the practice of forestry with the object of developing and maintaining forests of high scenic value. Recreational forests are being developed near towns and cities. The areas are being planted with flowering trees, shrubs and creepers to provide forest atmosphere near towns and cities.

AGROFORESTRY

Agroforestry is the management and integration of trees, crops and/or livestock on the same plot of land and can be an integral component of productive agriculture. It may include existing native forests and forests established by landholders. It is a flexible concept, involving both small and large-sized land holdings.

Scientifically speaking, agroforestry is derived from ecology and is one of the three principal land-use sciences, the other two being agriculture and forestry. Agroforestry differs from the latter two principals by placing an emphasis on integration of and interactions among a combination of elements rather than just focussing on each element individually.

Agroforestry has a lot in common with intercropping (the practice of planting two or more crops on the same plot) with both practices placing an emphasis on interaction between different plant species. Generally speaking, both agrofrestry and intercropping can result in higher overall yields and reduced operational costs.

Benefits of Agroforestry

Over the past two decades, a number of studies have been carried out analysing the viability of agroforestry. The combined research has highlighted that agroforestry can reap substantial benefits both economically and environmentally, producing more output and proving to be more sustainable than forestry or agricultural monocultures. Agroforestry systems have already been adopted in many parts of the world.

Agroforestry systems can include the following benefits:

- They can control runoff and soil erosion, thereby reducing losses of water, soil material, organic matter and nutrients.
- They can maintain soil organic matter and biological activity at levels satisfactory for soil fertility. This depends on an adequate proportion of trees in the system- normally at least 20% crown cover of trees to maintain organic matter over systems as a whole.
- They can maintain more favourable soil physical properties than agriculture, through organic matter maintenance and the effects of tree roots.

- They can lead to more closed nutrient cycling than agriculture and hence to more efficient use of nutrients. This is true to an impressive degree for forest garden/farming systems.
- They can check the development of soil toxicities, or reduce exiting toxicities-both soil acidification and salinization can be checked and trees can be employed in the reclamation of polluted soils.
- They utilize solar energy more efficiently than monocultural systems different height plants, leaf shapes and alignments all contribute.
- They can lead to reduced insect pests and associated diseases.
- They can be employed to reclaim eroded and degraded land.
- Agro forestry can augment soil water availability to land use systems. In dry regions, though, competition between trees and crops is a major problem.
- Nitrogen-fixing trees and shrubs can substantially increase nitrogen inputs to agro forestry systems.
- Trees can probably increase nutrient inputs to agro forestry systems by retrieval from lower soil horizons and weathering rock.
- The decomposition of tree and pruning can substantially contribute to maintenance of soil fertility. The addition of high-quality tree prunings leads to large increase in crop yields.
- The release of nutrients from the decomposition of tree residues can be synchronized with the requirements for nutrient uptake of associated crops. While different trees and crops will all have different requirement, and there will always be some imbalance, the addition of high quality prunings to the soil at the time of crop planting usually leads to a good degree of synchrony between nutrient release and demand.
- In the maintenance of soil fertility under agro forestry, the role of roots is at least as important as that of above-ground biomass.
- Agro forestry can provide a more diverse farm economy and stimulate the whole rural economy, leading to more stable farms and communities. Economics risks are reduced when systems produce multiple products.

As well as building on practices used in forestry and agriculture, agroforestry also works towards land protection and conservation through more effective protection of stock, control of soil erosion, salinity and water tables and a higher quality control of timber.

A denser, more-dependable tree covering can provide shelter to livestock during the warmer months allowing the animals can conserve energy. That same tree covering helps block out wind, helping to boost water retention levels that can help produce a more robust crop yield.

Agroforestry can improve land protection in the following areas:

- Salinity and water table control: Salinity is mainly caused by rising water tables. Trees help to lower water tables, acting as pumps to take up water from the soil and then evaporating it to the atmosphere.
- Soil erosion control: Soil erosion or loss results from the action of wind and water on unprotected soils. The forest canopy, roots and leaf litter all have a role in controlling soil erosion.
- Water logging: Through water removal, established trees can substantially reduce water logging in their immediate area, which may result in improved land uses, e.g. pasture or crop. Agroforestry can have immense benefits for the environment and the farmer. For farmers, the ability to maintain some sort of control over land and production in the face of climate change means agrofrestry could hold huge promise for the agricultural sector. On an environmental level, agroforestry's ability to help prevent soil erosion while simultaneously aiding water retention and promoting soil fertility could help provide a solution for areas where rainfal is irregular or might become irregular due to climate change while dense plantations of trees would also help absorb CO_2 and regulate local temperature.

In agroforestry systems, woody and herbaceous perennials are grown on land that also supports agricultural crops or animals. The mixture of these components, in the form of spatial arrangement or temporal sequence, enhances ecological stability and production sustainability. This integration allows the components to complement one another in their use of resources and in the timing of that use. Perennials have deeper roots and higher canopies than those of annuals, allowing better management of above- and below-ground resources. Under ideal conditions:

- Nutrients recycled from the subsoil to the surface by deep-rooted perennials can be used by annuals.
- Leguminous perennials fix atmospheric nitrogen that can be used by annuals.
- There is minimal competition for water because of differences in depth from which the roots of annuals and perennials extract water from the soil.
- Some perennials produce allelopathic compounds that can suppress weeds.
- Differences in the structure of perennials and annuals, leading to a multistory canopy, reduce competition for light among plants.

Agroforestry systems have the potential to improve production and to enhance the agronomic and ecological sustainability of resource-poor farmers in the humid tropics. In practice, however, the potential benefits of agroforestry systems can be harnessed only through skillful and labor-intensive management of compatible systems. There are no simple blueprints of a universally applicable system that can harness all potential benefits possible under ideal conditions. Thus, a wide range of agroforestry systems has been designed to alleviate agronomic, ecologic, or managerial constraints.

Types of Traditional Agroforestry Systems in the Humid Tropics

Agroforestry is not a new concept in the humid tropics. Several types of traditional agroforestry systems exist, but no standard classification system is available to categorize them. Nair proposed a classification system based on structural, functional, agroecological, and socioeconomic factors. These broad categories are interrelated, and not necessarily mutually exclusive. In agroforestry land use systems, three basic components are managed by people: The tree (woody perennial), the herb (agricultural crops, including pasture species), and the animal. Based on their structure and function, agroforestry systems can be classified into the following three categories:

- Agrisilviculture is the use of crops and trees, including shrubs or vines. It includes shifting cultivation, forest gardens, multipurpose trees and shrubs on farmland, alley cropping, and windbreaks as well as integrated multistory mixtures of plantation crops.
- Silvopastoral systems are combinations of pastures (with or with out animals) and trees. They include cut-and-carry fodder production, living fences of fodder trees and hedges, and trees and shrub grown on pastureland.

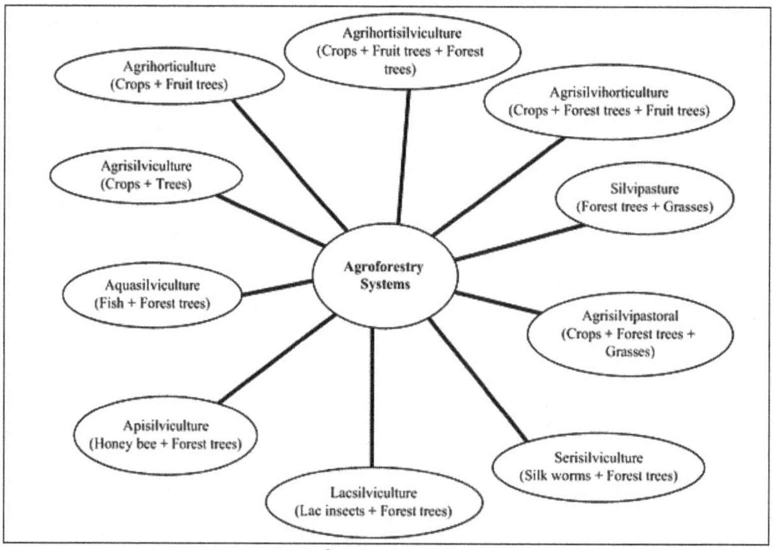

Agroforestry systems.

- Agrisilvopastoral systems are those that combine food crops, pastures (with or without animals), and trees and include home gardens and woody hedges used to provide browse, mulch, green manure, erosion control, and riverbank stabilization.

Other types of agroforestry systems include apiculture (beekeeping) using honey-producing trees, aquaculture whereby trees lining fishponds provide leaves as forage for fish, and multipurpose woodlots that serve various purposes such as wood, fodder, or food production and soil protection or reclamation.

Principal types of agrisilvicultural systems traditionally used in the humid tropics are:

- Rotational agroforestry: In traditional shifting cultivation, trees and wood species are naturally regenerated over a period of 5 to 40 years and rotated with annual crops that are cultivated from 1 to years. Improved tree species can be grown in place of native vegetation to achieve better soil conditions. This technique is used in multipurpose woodlots (where diverse mixtures of trees are used), home gardens (where trees and crops are grown close to the house), and compound farms (where trees, animals, crops, and the farmer's dwelling are in a fenced area).
- An intercropping system: Annual and perennial groups of plants are grown within the same land management unit. This system enables continuous production of food and tree products with a minimum need for restorative or idle fallow. Typical examples of intercropping systems include alley cropping and boundary planting of trees and wood hedges. Two examples of the successful use of agroforestry systems by resource-poor farmers in the tropics are found in the Philippines and Rwanda. In the Philippines, many small-scale farmers took up cash-crop-tree farming to produce pulpwood, poles, timber, charcoal, or fuelwood in the 1960s. The program gained significant momentum in 1972 when the Paper Industries Corporation of the Philippines (PICOP) entered into an agreement with the Development Bank of the Philippines to develop a loan scheme for small-scale tree farmers with titled or untitled land. Provision was made for part of the farm area to be maintained under food crops. PICOP guaranteed a minimum purchase price, but allowed farmers to sell wood to other outlets if they could get better prices. Within 10 years, the program covered 22,000 ha and supported 3,800 farmers, about 30 percent of whom had taken advantage of the credit program. A key to the program's success has been the high financial returns from tree growing. Adequate market incentives and security of land tenure were the basic factors responsible for acceptance by farmers.

The second example involves restoring eroded land in Rwanda using an agrisilvopastoral system. At Nyabisindu, a complex system of trees, animals, and crops was developed using the community's existing knowledge. Trees and hedges, yielding fruit, wood, and fodder, were used as protective ground cover against soil erosion. Extensive use was also made of perennial crops to further stabilize the soil.

In Amazonian Ecuador, a sustainable system has been developed to raise sheep in association with cassava and contour strips of Inga edulis, which is a deep-rooted leguminous fuelwood tree. After the cassava is harvested, a perennial leguminous ground cover, Desmodium sp., is planted between the trees to enrich the soil. Sheep graze on the ground cover.

Keys to the success of these projects included building on traditional knowledge, involving farmers in the choice of species, and providing economic incentives greater than those of traditional systems. Resource conservation and land restoration were additional benefits to the local community.

The viability and sustainability of these systems can be attributed to some combination of the following factors:

- A reduced fallow period and a greater ability to cultivate on a long-term basis, thereby eliminating the need to move to new land.
- Reduced use of chemical fertilizers and other fossil-fuel-based inputs due to enhancement of soil organic matter and improvement in soil fertility.
- Improved soil structure and physical properties (for example, better sizes of pores and channels in the soil that allow better water penetration and drainage).
- Decreased risks of soil degradation from accelerated erosion and other degenerative processes.
- Increased production and a rise in economic status from subsistence to partially commercialized farm.
- Decreased need for clearing new land.
- Improved Agroforestry Systems.
- Scientists and policymakers generally are eager to improve traditional agroforestry systems by enhancing productivity and ecological compatibility. Ways of improving these systems include using better trees and woody shrubs and creating an orderly arrangement of trees, crops, and livestock.

Arrangement of Trees, Crops and Livestock

Rather than using a random and difficult-to-mechanize system of growing trees with annuals or animals, mixtures can be grown in an improved spatial or temporal arrangement. In an agrisilvicultural system, for example, trees can be grown in alternate rows or strips, as contour hedges to control erosion, or on field boundaries. These orderly arrangements can facilitate the use of animal power and of mechanization of farm operations, save labor, and enhance economic and ecological benefits.

Alley cropping is a common example of a spatial arrangement. Food crops are grown in alleys formed by contour hedgerows of trees or shrubs. Trees and shrubs can be pruned to prevent shading of the food crops and to provide nitrogen-rich mulch for crops and fodder for livestock. Shrubs and trees also act as windbreaks, facilitate nutrient recycling, suppress weed growth, decrease runoff, and reduce soil erosion.

The most common trees for alley cropping are fast-growing, multipurpose, nitrogen-fixing trees. Tree species with the potential for use with nonacid tropical soils include Acioa baterii, Alchornea cordifolia, Gliricidia septum, and Leucaena leucocephala. Species for acid soils include Acioa baterii, Alchornea cordifolia, Anthonotha macrophylla, Calliandra calothyrsus, Cnestis ferruginea, Dialium guineense, Erythrina spp., Flemingia congesta, Harungana madagascariensis, Inga edulis, Nuclea latifolia, and Samanea saman. Hedgerows of Cassia spp., G. septum, and L. Ieucocephala can be established from seed. Other species are established from seedlings or stem cuttings. However, the use of stem cuttings often results in a patchy stand with a high rate of mortality. Trees established from stem cuttings are also easily uprooted because of poor root system development.

Alley cropping systems can produce two or more products, such as food grains, fodder, mulch, fuelwood, and staking and building materials, and can increase or maintain soil structure. However, the beneficial effects of these systems depend on many factors, such as the tree species, area of land allocated to trees, hedgerow management, crop management, soil type, and prevalent climate. In areas with nonacid soils, satisfactory yields of cereals can be attained with the added benefit of erosion control. These systems are also labor intensive, therefore they are adapted primarily to areas of high population density and modest to low labor cost.

Advantages and Disadvantages of Agroforestry

Given a compatible association of trees and annual crops, agroforestry systems are likely to sustain economic productivity without causing severe degradation of the environment. Because of the low fertility of most upland tropical soils, some degradation is inevitable with any cultivation system. The rate and risks of such degradation are lower with agroforestry than with annual crop rotations. Soil organic matter, pH, soil structure, infiltration rate, cation exchange capacity, and the base saturation percentage are maintained at more favorable levels in agroforestry systems due to reduced losses to runoff and soil erosion, efficient nutrient recycling, biological nitrogen fixation by leguminous trees, favorable soil temperature regime, prevention of permanent changes in soil characteristics caused by drying, and improved drainage because of roots and other biochannels.

It is important to note, however, that trees have both positive and negative effects on soils. Negative effects include growth suppression caused by competition for limited resources (nutrients, water, and light) and by allelopathic effects. Mismanagement of trees (through, for example, improper fertilizer application or inadequate water control) can also cause soil erosion, nutrient depletion, water logging, drought stress, and soil compaction.

Economic evaluation is an important tool to assess a technology. Labor-intensive alley cropping can be economical under severe cash constraints and where hired labor is available at relatively low cost. The available data on alley cropping indicate that the system cannot sustain production without supplemental inputs of chemical fertilizers if high yields are desired. In fact, soil degradation and attendant yield reductions can occur even with the fertilizer application.

Erosion control is a definite advantage of closely spaced contour hedgerows of L. Ieucocephala or other shrubs, but it can also be achieved through cover crops, grass strips, or no tillage. Nonetheless, the erosion preventive effects of L. Ieucocephala hedgerows must also be considered in evaluating the economic impact of an alley cropping system.

Data on soil properties indicate that intensive cultivation resulted in decreases in soil organic matter content, total nitrogen, pH, and exchangeable calcium, magnesium, and potassium in all systems including alley cropping and control. This drastic decline in soil fertility was observed in relatively fertile soils. The relative rates of decline, however, were somewhat less in alley cropping than with prow-based control. These results are also supported by data on acidic tropical soils in Yurimaguas, Peru. Szott observed significantly more calcium, magnesium, phosphorus, and potassium in the upper 15 cm of soil with control without trees treatments than with alley-cropping treatments. Fertilized control without trees significantly exceeded all other treatments in topsoil calcium and magnesium. The pH values were also significantly greater in the fertilized control.

Priorities

The agronomic aspects and biophysical processes of agroforestry using traditional cropping systems need to be more fully evaluated. For example, farmers using traditional systems commonly space their plants more widely apart than farmers using improved systems, and hence grow fewer plants per unit area. More scientific data are needed on interactions among plant species, specifically in relation to competition for water, nutrients, and light, and on the suppression of growth of one species by another species' release of toxic substances.

Major distinctions also should be made for research on acidic versus nonacidic soils. Too often soils and their constraints are ignored when designing or evaluating agroforestry systems. The ability of agroforestry systems to enhance nutrient availability on infertile soils is very limited compared with systems on fertile soils. On both, however, agroforestry systems can play an important role in reducing nutrient losses. Although litter production and quantities of nutrients recycled in litter are greater on fertile than on infertile soils, management techniques for accelerating nutrient fluxes through pruning hold promise for increasing plant productivity on infertile soils. More information is needed on the magnitude of and controls on belowground litter production and how it can be managed. Litter decomposition and soil organic matter dynamics in agroforestry systems might most easily be manipulated by managing woody vegetation to produce organic residues of a certain quality and to regulate soil temperature and moisture. More attention needs to be paid to specific soil organic matter pools, their importance in nutrient supply and soil structure, how they are affected by soil properties, and how they can be managed.

In addition to understanding the agronomic and biophysical aspects of agroforestry systems, the social, ecological, and economic elements require more attention. The economic feasibility of agroforestry systems needs to be assessed at the farm level. Human ecology and sociology play an important role in the acceptance and spread of technologies, as do the specific sociopolitical and institutional constraints.

Agroforestry can be a sustainable alternative to shifting cultivation. However, systems suited to many major soils and ecological regions of the tropics have yet to be developed. For example, alley cropping has shown some advantages in Alfisols but not in other soils and harsh environments. Further research is needed to develop systems performance indicators and to document ecological viability of agroforestry systems across a range of biophysical conditions.

Types of agroforestry systems:

- Structural basis.
- Nature of components.

Agrisilvicultural Systems

In this system, agricultural crops are intercropped with tree crops in the interspace between the trees. Under this system agricultural crops can be grown upto two years under protective irrigated condition and under rainfed farming upto four years. The crops can be grown profitably upto the above said period beyond which it is uneconomical to grow grain crops. However fodder crops, shade loving crops and shallow rooted crops can be grown economically. Wider spacing is adopted without sacrificing tree population for easy cultural operation and to get more sunlight to the intercrop. Performance of the tree crops is better in this system when compared to monoculture.

Silvopastoral Systems

The production of woody plants combined with pasture is referred to Silvipasture system. The trees and shrubs may be used primarily to produce fodder for livestock or they may be grown for timber, fuelwood, fruit or to improve the soil. This system is classified in to three categories.

- Protein bank: In this Silvipastoral system, various multipurpose trees (protein rich trees) are planted in or around farmlands and range lands for cut and carry fodder production to meet the feed requirement of livestock during the fodder deficit period in winter. Example: Acacia nilotica, Albizia lebbeck, Azadirachta indica, Leucaena leucocephala, Gliricidia sepium, Sesbania grandiflora.

- Livefence of fodder trees and hedges: In this system, various fodder trees and hedges are planted as live fence to protect the property from stray animals or other biotic influences. Example: Gliricidia sepium, Sesbania grandiflora, Erythrina sp, Acacia sp.

- Trees and shrubs on pasture: In this system, various tree and shrub species are scattered irregularly or arranged according to some systemic pattern to supplement forage production. Example: Acacia nilotica, Acacia leucophloea ,Tamarindus indica, Azadirachta indica.

Agrosilvopastoral Systems

The production of woody perennials combined with annuals and pastures is referred Agrisilvopastural system.

This system is grouped into two categories:

Home Gardens

This system is found extensively in high rainfall areas in tropical South and South east Asia. This practice finds expression in the states of Kerala and Tamil Nadu with humid tropical climates where coconut is the main crop. Many species of trees, bushes, vegetables and other herbaceous plants are grown in dense and in random or spatial and temporal arrangements. Most home gardens also support a variety of animals. Fodder grass and legumes are also grown to meet the fodder requirement of cattle.

Home gardens represent land use systems involving deliberate management of multipurpose trees and shrubs in intimate association with annual and perennial agricultural crops and livestock within the compounds of individual houses. The whole tree-crop- animal units are being intensively managed by family labour. Home gardens can also be called as Multitier system or Multitier cropping.

Home gardens are highly productive, sustainable and very practicable. Food production is primary function of most home gardens.

Structure of Home Gardens

Home gardens are characterized by high species diversity and usually 3-4 vertical canopy strata. The layered configuration and compatible species admixture are the most conspicuous characteristics of all home gardens. Generally all home gardens consist of an herbaceous layer near the ground, a tree layer at the upper levels and an intermediate layer. The lower layer can be partitioned into two, the lowermost being at less than 1.0m in height, dominated by different vegetables and the second layer of 1.0 -3.0/m height comprising food crops such as banana, papaya and so on. The upper tree layer can also be divided into two, consisting of emergent , full grown timber and fruit trees occupying the upper most layer of 25m height and medium size trees of 10-20m occupying the next lower layer. The intermediate layer of 5-10m height is dominated by various fruit trees.

Choice of Species:

- Woody species: Anacardium occidentale,Artocarpus heterophyllus, Citrus spp, Psiduim guajava, Mangifera indica, Azadirachta indica, Cocus nucifera.
- Herbaceous species: Bhendi, Onion, cabbage, Pumpkin, Sweet potato, Banana, Beans, etc.

Woody Hedgerows

In this system various woody hedges, especially fast growing and coppicing fodder shrubs and trees are planted for the purpose of browse, mulch, green manure, soil conservation etc. The following species viz., Erythrina sp, Leucaena luecocephala, Sesbania grandiflora are generally used.

Other Systems

- Apiculture with trees: In this system various honey (nector) producing trees frequently visited by honeybees are planted on the boundary of the agricultural fields.

- Aquaforestry: In this system various trees and shrubs preferred by fish are planted on the boundary and around fish ponds. Tree leaves are used as feed for fish. The main role of this system is fish production and bund stabilization around fish ponds.
- Mixed wood lots: In this system, special location specific MultiPurpose Trees (MPTs) are grown mixed or separately planted for various purposes such as wood, fodder, soil conservation , soil reclamation etc.

Arrangement of Components

- Spatial Arrangement: Spatial arrangement of plants in an agroforestry mixture may result in dense mixed stands (as in home gardens) or in sparse mixed stands (as in most systems of trees in pastures).
- Temporal Arrangement: Temporal arrangements of plants in Agroforestry may also take various forms. An extreme example is the conventional shifting cultivation cycles involving 2-4 years of cropping and more than 15 years of fallow cycle, when a selected woody species or mixtures of species may be planted. Similarly, some silvipastoral systems may involve grass leys in rotation with some species of grass remaining on the land for several years. These temporal arrangement of components in agroforestry are termed coincident, concomitant, overlapping, separate and interpolated.

Functional Basis

All agroforestry systems have two functions:

Productive Functions

The Productive functions are:

- Food.
- Fodder.
- Fuel wood.
- Cloths.
- Shelter.
- NTFPs.

Protective Functions

The Protective functions are:

- Wind breaks.
- Shelterbelts.
- Soil conservation.
- Soil improvement.

Socio-economic Classification

Based on socioeconomic criteria as scale of production and level of technology input and management, agroforestry systems have been grouped in to three categories:

Commercial AF Systems

The term commercial is used whenever the scale of the production of the output is the major aim of the system. Examples:

- Commercial production of plantation crops such as rubber, oilpalm, and coconut with permanent underplanting of food crops, pastures.
- Commercial production shade tolerating plantation crops such as coffee, tea and cocoa under overstorey of shade trees.

Intermediate AF Systems

Intermediate systems are those between commercial and subsistence scale of production and management. Examples: Production of perennial cash crops and subsistence food crops undertaken on farms wherein the cash crops fulfill the cash needs and the food crops meet the family's food needs.

Subsistence AF Systems

Subsistence AF systems are those wherein the use of land is directed towards satisfying basic needs and is managed mostly by the owner and his family.

Ecological Classification

- Agroforestry systems in Humid\Subhumid lowlands Examples: Homegardens, Trees on rangelands and pastures, improved fallow in shifting cultivation and Multipurpose woodlots.
- Agroforestry systems in Semiarid and arid lands Examples: Various forms of silvopastoral systems, wind breaks and shelterbelts.
- Agroforestry systems in Tropical High lands Examples: Production systems

involving plantation crops such as coffee, tea, use of woody perennials in soil conservation and improved fallow.

Agroforestry System Components

Agroforestry systems can be made up of many different woody components (as well as the livestock and crop components).

- Alley cropping - Rows of trees (either standards or coppice) in between alleys of crops or pasture.
- Hedges.
- Shelter belts.
- Trees and woodland round riparian zones.
- All types of woodland areas – coppice woodland, woodland with standards and coppice woodlands with standards.
- Orchards.
- Grazed woodland.
- Trees in pasture.
- Pollards.

In an agroforestry system all woody components should be fully integrated into the agricultural production system. The woody components should have a beneficial ecological impact on the land and other system components (i.e. livestock and crops). Through management, it should also be possible to gain an economic profit from the woody components either directly by selling products, or indirectly from the woody components having a beneficial effect on other system components (i.e. the shelterbelt producing higher yielding crops or sheep).

For an Eco-Agroforestry approach, diversity should exist on all levels of the system:

- Within species diversity of trees, crops and livestock.
- Diversity of tree, crops and livestock species.
- Diversity of system components at the landscape level.

Benefits

Trees in agroforestry systems modify microclimatic conditions including temperature, water vapour content of air and wind speed, which can have beneficial effects on crop growth and animal welfare. Additionally a wide range of other services can be provided including:

- Soil management.
- Weed control.
- Natural fencing.
- Carbon sequestration.
- Nutrient recycling.

By minimising nutrient losses and maximising internal cycling of nutrients, and by enhancing pest and disease control, agroforestry systems reduce the need for agrochemical inputs. Agroforestry has a further role in protecting the environment and providing a number of ecosystem services, such benefits include:

- Regulation of soil, water and air quality.
- Enhancement of biodiversity.
- Climate change mitigation and adaptation.

Products

Agroforestry systems support the production of a wide range of products including:

- Food.
- Fuel.
- Timber.
- Fodder and forage.
- Fibre.
- Gums and resins.
- Thatching and hedging materials.
- Gardening materials.
- Medicinal products.
- Craft products.

- Recreation.
- Ecological services.

The stack of photos shows a short rotation coppice/potato silvorable system at Wakelyns agroforestry. The short rotation hazel coppice is for energy, from field to wood chip boiler.

These products can impact the local economy through increasing:

- Economic stability.
- Diversification of local products and economies.
- Diversification of rural skills.
- Improved food and fuel security.
- Improvements to the cultural and natural environment.
- Landscape diversification.

Scope of Agroforestry

Agroforestry applies to private agricultural and forest lands and communities. These are highly disturbed, human-dominated land-use systems. Targets include highly-erodible, flood-prone, economically marginal and environmentally sensitive lands. The typical situation is agricultural, where trees are added to create desired benefits. Our goal is to restore essential processes needed for ecosystem health and sustainability, rather than to restore natural ecosystems. Agroforestry provides strong incentives for adoption of conservation practices and alternative land uses, and supports a collaborative watershed analysis approach to management of landscapes containing mixed ownerships, vegetation types and land uses.

Bamboo plantation under agroforestry system.

Possible Impacts of Agroforestry

- Controlling poverty through increased income by higher production of agroforestry products for home consumption and market. Bamboo plantation under agroforestry system.

- Food security by restoring farm soil fertility for food crops and production of fruits, vegetables, nuts and edible oils.
- Empowerment to women farmers and other less-advantaged rural residents whose rights to land are insecure through better negotiations.
- Reducing deforestation and pressure on forest by providing fuelwood grown on farms.
- Increasing buffering capacity of farmers against the effects of global climate change onfarm tree crops and tree cover.
- Improving soil health of the farm through ameliorated micro-climate and nutrition level.
- Augmenting accessibility to medicinal trees for cure of common and complex diseases.

Features of Agroforestry

Agroforestry practices are intentional systematic combinations of trees with crops and/or livestock that involve intensive management of the interactions between the components as an integrated agro ecosystem. These key features are the essence of agroforestry and are what distinguish it from other farming or forestry practices. To be called agroforestry, a land-use practice must satisfy following criteria:

- Intentional: Combinations of trees, crops and animals are intentionally designed and managed as a whole unit, rather than as individual elements that may occur in close proximity but are controlled separately.
- Intensive: Agroforestry practices are intensively managed to maintain their productive and protective functions; these practices often involve annual operations such as cultivation and fertilization.

- Interactive: Agroforestry management seeks to actively manipulate the biological and physical interactions between the tree, crop and animal components. The goal is to enhance the production of more than one harvestable component at a time, while also providing conservation benefits such as non-point source water pollution control or wildlife habitat.
- Integrated: The tree, crop and animal components are structurally and functionally combined into a single, integrated management unit. Integration may be horizontal or vertical, and above or below ground. Such integration utilizes more of the productive capacity of the land and helps balance economic production with resource conservation.

Significance of Agroforestry

Agroforestry provides a different land use option, compared with traditional arable and forestry systems. It makes use of the complimentarily relationship between trees and crops, so that the available resources can be effectively utilized. It is a practice that supports the environment and has an obvious landscape benefit. Efficient, modern versions of agroforestry have been developed, that are adapted to the constraints imposed by mechanization. The agroforestry plot remains productive for the farmer and generates continuous revenue, which is not feasible in arable land. Agroforestry allows for the diversification of farm activities and makes better use of environmental resources. Owing to increase in population of human and cattle, there is increasing demand of food as well as fodder, particularly in developing countries like India. Each year farmers of the world must now attempt to feed 81 million more people irrespective of weather. It is important to note that there will be 19% decline in cropland per head by the end of this century due to population explosion. Therefore, there is slight scope to increase food production by increasing the area under cultivation. A management system therefore, needs to be devised that is capable of producing food from marginal agricultural land and is also capable of maintaining and improving quality of producing environment.

Agroforestry has interesting advantages from three different perspectives.

Agriculture Perspective

- Diversification of the activities of arable farmers, with the building-up of an inheritance of multi-purpose trees, with continuous revenue from farm.
- Protection of intercrops and animals by the trees, which have a windbreak effect, providing shelter from the sun, rain, wind, soil erosion and stimulating soil microfauna and microflora.
- Recycling of some of the leached or drained nutrients by the deep roots of the trees; enrichment of the soil organic matter by tree litter and by the residue of the trees.

- Possibility of combining the interest of the farmer (for an inheritance of wood) and the farm (for access to cultivated land). Possible increased remuneration for the arable farmer for the trees.
- An alternative to full reforestation of arable land, permitting the continuation of arable activity on land whose arable potential otherwise is conserved. The tree component can be reversed, the plot stays "clean" (free from scrub) and is easy to destump when the trees are clear felled (the stumps are in lines and few in number).
- In silvipastoral plots, fodder units can be available at different periods compared to full cropped plots, extending the grazing calender.

Forestry Perspective

- Enhancement in the biomass growth of the trees by wide spacing (+80% over 6 years in the majority of the experimental plantations). A large reduction in the maintenance costs of the plantation, due to the presence of the intercrops.
- Improvement in the quality of wood produced (wide regular rings, suited to the needs of industry), because the trees are not subjected to cycles of competition and thinning.
- Guaranteed follow-up and tree care due to the arable intercropping activity. In particular, protection against the risk of fire in susceptible areas, with pasture or with intercrops like vine or winter cereals (clear bare ground in summer after stubble ploughing).
- Agroforestry plantations on arable land allow the development of a quality wood resource that complements, rather then competes with, the products from traditionally exploited forests. It is especially important to produce wood that can substitute for tropical saw logs, which will soon decline in availability and quality. The areas concerned will remain small in terms of their absolute value, but the production of wood from them could become a critical input to the European wood supply network. Tree species that are little used in forestry,

but are of high value, could be grown in agroforestry systems: service trees, pear trees, common sorbs, walnut trees, wild cherry trees, maple trees, tulip trees, paulownias, etc.

From the Environmental Perspective

- Improvement to the development of natural resources: the total wood and arable production from an agroforestry plot is greater than the separate production obtained by an arable-forest separate cropping pattern on the same area of land. This effect results from the stimulation of complementarity between trees and crops on agroforestry plots. Thus, weeds, which are spontaneously present in young forestry plantations, are replaced by harvested crops or pasture; maintenance is less costly and environmental resources are better used.

- Better control of cultivated areas of land: by substituting for arable plots, the agroforestry plots contribute to diminishing the cultivated area of land. The intensification of environmental resource use by agroforestry systems is not resulting in more crop products.
- Creation of original landscapes that are attractive, open and favor recreational activities. Agroforestry plots have a truly innovative landscaping potential, and would improve the public image of farmers to society. This will be particularly the case in very sparsely wooded areas, where plots are developed by planting arable land, and in very heavily wooded areas, where plots are developed by thinning the existing forest.
- Counteract the greenhouse effect: constitution of an effective system for carbon sequestration, by combining the maintenance of the stock of organic material in the soil (the case especially with meadows), and the superimposition of a net fixing wooded layer.
- Protection of soil and water, in particular in sensitive areas.
- Improvement of biodiversity, especially by the abundance of "edge effects". This in particular, permits a synergistic impro integrated protection of crops by their association with trees, chosen to stimulate the hyperparasite (parasites of parasites) population of crops, is a promising way forwards.

Young walnut trees in lavanda plot.

- These favorable characteristics are vement, by favoring the habitat of game. The 7. as coherent with the many objectives of the laws guiding agriculture and forestry, as they are with the directing principles of the Common Agricultural Policy.

Benefits from Agroforestry

Environment Benefits

Combining trees with food crops on cropland farms yield certain important environment benefits, both general ecological benefits and specific on-site benefits.

The general ecological benefits include:

- Reduction of pressure on forest.
- More efficient recycling of nutria.
- Better protection of ecological systems.
- Reduction of surface run-off, nutrient le effect of tree roots and stems of these processes.
- Improvement of microclimate, such as lower in reduction of evaporation of soil moisture through a combination of mulching and shading.
- Increment in soil nutrients through addition and decomposition of litter-fall.
- Improvement of soil structure through the constant addition of organic ma decomposed litter.

Economic Benefits

Agroforestry systems on croplands/farmlands bring significant economic benefits to the farmer, the community, the region or the nation. Such benefits may include:

- Increment in an maintenance of outputs of food, fuelwood, fodder, fertilize timber.

- Reduction in incidence of total crop failure, common to single-cropping or monoculture system.
- Increase in levels of farm incomes due to improved and sustained productivity.

Social Benefits

Besides the economics benefits, social benefits occur from increase in crop and tree product yields and in the sustainability of these products. These benefits include:

- Improvement in rural living standards from sustained employment and higher incomes.
- Improvement in nutrition and health due to increased quality and diversity of food outputs.

Importance of Agroforestry

- Improved nutrition and food security: Planting trees in agricultural lands can help increase food production and boost food security. Moreover, agroforestry helps in health and nutrition improvement due to increased diversity and quality of food outputs. Tree planting, as part of agroforestry, can be used as a source of fuel, food and non-wood products that can be consumed or sold resulting in additional food and security.

  Generally, trees can provide nutritious fruits, nuts, and leaves for consumption in households. Felled trees and their residues can be used as wood energy for cooking and heating while leaves can be used as forage for livestock. In addition to food products, agroforestry also supports the production of a wide range of products such as timber, fibre, fodder and forage, craft products, medicinal products, hedging materials, and gums and resins among others.

- More and more benefits from Trees: Planting trees, as part of agroforestry practice, can be beneficial for the health and survival of humans, animals, and crops. Trees can block strong winds, protecting crops from damage. Some crops like barley, alfalfa, and winter wheat are also known to thrive under shelter. Combining proper tree species and responsive crops can lead to improved yields.

  Trees also protect animals from wind chill in cold days while also providing shade on hot days and thus, helping lower animal stress. Furthermore, changes in temperature can cause irregular feeding habits and hence, growing trees can help increase feeding efficiencies. Trees can also be a source of medicines and natural remedies.

- Creation of resilient livelihoods: Growing trees together with crops and animals can help reduce the vulnerabilities associated with agricultural production and even improve the recovery after natural disasters, hazards, or socioeconomic downturns. Agroforestry economics, for example, helps to increase the diversity of production within the system thereby reducing the risk of economic failure.

The roots of trees can strengthen the soil structure; mitigating soil erosions, improving soil fertility, and preventing possible landslides. Furthermore, trees can help prevent desertification and its social, agricultural, and environmental consequences. As such, improved and sustainable productivity leads to increases in levels of farms income. It is also worth noting that agroforestry can bring forth sustained employment and higher income, which lead to an improvement in rural living standards.

- Climate change mitigation and adaptation: Growing trees in agricultural systems can reduce the impact of climate change on agriculture. Sourcing tree products from trees grown on farms reduces the need to cut trees and hence reduce the rate of deforestation that is quickly getting rid of the planets carbon sinks while releasing stored carbon into the environment. Trees in agroforestry systems, therefore, help to mitigate climate change through carbon sequestration.

  Further, Agroforestry plays a critical role in improving climatic conditions and weather patterns by modifying microclimatic conditions such as temperature and water vapor content of air and wind speed, resulting in beneficial effects on crop growth and animal welfare. This in general, can limit the effects of climate change and global warming on agriculture as it can help in regulating air quality, water concentration, rain cycles, and patterns, and wind erosion.

- Environmental benefits: Careful planning and integration of agroforestry can help protect the natural resources in the environment. For example, growing trees can help improve the quality of water and its quantity by filtering and capturing of water resources. Trees also support biodiversity by providing a suitable environment for insects, animals, and plants. If trees with nitrogen-fixation function are part of the system, agroforestry can help restore soil fertility.

  Other environmental benefits of agroforestry systems include reduction of pressure on natural forests, more efficient recycling of nutrients by deep-rooted trees on the site, increment in soil nutrients, improvement of microclimate and reduction of surface run-off. The systems also help improve soil structure and provide better protection of ecological systems.

- Support local communities and cultures: Agroforestry is designed to help local communities and cultures thrive. With the help of agroforestry specialists, indigenous people and local communities can continue with the local beliefs and culture while ensuring long term sustainability of the traditional systems. Furthermore, by preserving indigenous working techniques and species, agroforestry also helps to protect humankind's agricultural heritage. It is also worth noting that agroforestry can lead to decent rural livelihoods, cultural diversity and maintaining of local spiritual beliefs. Agroforestry also helps in the stabilization and improvement of local communities by eliminating the need to move sites of farm activities.

- Can reduce poverty in some areas if practiced sustainably: Trees and tree products have economic value that can get agroforesters a source of livelihood and potentially reduce their poverty levels, especially in developing or emerging economies. The value addition of newly-produced tree products can be a source of employment and income for individuals.

  Furthermore, farmers can take advantage of the incentives offered to support agroforestry as sources of income. It is also worth noting that growing trees can help reduce the production costs resulting in increased household income. In addition to increment in outputs of food, fodder, fuelwood, and timber, agroforestry systems also help in the reduction in incidences of total crop failure, which is common in monoculture and single cropping systems.

Components of Agroforestry

In modern agroforestry, the systems are classified based on the present components of its practice. For example, where trees are grown together with crops, the system is referred to as silvoarable whereas silvopastoral is where trees are grown together with animals. Agro-silvopastoral is where plants, animals, and trees are kept together.

It is also becoming increasingly common to find specialized systems like the entomoforestry where trees and insects like silk months and honey bees are the main components and aquaforestry where trees and fish are the main components. Depending with the system, some of the woody components you will find include shelter belts, hedges, alley cropping, orchards, grazed woodland, and pollards among others. Other components include livestock and crop components.

- Agrisilvicultural systems: In this system, the components are agricultural crops and tree crops intercropped between the trees. The agricultural crops in this system can be grown for up to two years under protective irrigation and up to four years under refined farming. Beyond the specified period, growing grain crops becomes uneconomical but other crop components like fodder crops, shallow-rooted crops, and shade-loving crops can continue to be grown economically. To get the best yields, farmers should adopt wider spacing.

- Silvopastoral systems: In this system, the components are woody plants that are grown for pasture. The components can be trees or shrubs primarily grown to produce fodder for livestock or for fruit, timber, fuelwood or to improve the soil. This system can be classified into three categories that include protein bank, live fence of fodder trees and hedges, and hedges and shrubs on pasture.
- Protein Bank: In this category, the system components are multipurpose trees that are usually protein-rich trees planted in or around the farmlands for providing animal feeds. The common components in this system are Acacia nilotica, Azadirachta indica, Albizia lebbeck, Gliricidia sepium, Leucaena leucocephala, and Sesbania grandiflora.
- Live fence of fodder trees and hedges: In this category, hedges and fodder trees are planted as a live fence. The common components in this category are Erythrina sp, Sesbania grandiflora, Gliricidia sepium, and Acacia sp.
- Trees and shrubs on pasture: In this category, various shrub and tree species are planted irregularly or in a systematic pattern to supplement forage production. Some of the common components in this category are Acacia leucopheloea, Acacia nilotica, Azadirachta indica, and Tamarindus indica.
- Agrosilvopastoral systems: In this system, woody perennials are combined with annuals and pastures. The systems can be classified into two categories that include home gardens and woody hedgerows for browse, green manure, mulch, and soil conservation.
- Home gardens: This system is more common in areas with high rainfall areas in South and Southeast Asia where coconut is the main crop. Many tree species can be used in this system as well as vegetables, bushes, and herbaceous plants are grown randomly or in special or temporal arrangements. The system also supports a variety of animal components.

  Home gardens are more sustainable, highly productive and very practicable for food production. The woody components for home gardens are Artocarpus heterophyllus, Anacardium occidentale, cocus Nucifera, Psidium guajava, Azadirachta indica, Citrus spp, and Anacardium occidentale. The herbaceous components include Onion, Bhendi, pumpkin, cabbage, beans, banana, and sweet potato.
- Woody hedgerows: In this system, woody hedges especially the fast-growing,

are preferred for the purpose of mulch, browse, soil conservation and green manure. The common components include Leucaena luecocephala, Erythrina sp, viz, and sesbania Grandiflora.

- Other systems: Other systems such as *apiculture with trees*, *aquaforestry* and *mixed wood lots* also exist. In *apiculture system with trees*, the main components are honey bees and trees that encourage honey production. In the *aquaforestry system*, the main components are fish and trees and shrubs that are preferred by fish. In the system, the leaves are used as feed for fish while the trees are used for bund stabilization around the fish ponds. In *Mixed wood lots*, multipurpose trees are grown mixed or separately for various uses such as fodder, wood, soil reclamation, and soil conservation.

Challenges to Agroforestry

Although agroforestry systems can be advantageous, they are not widespread in the US as of 2013.

As suggested by a survey of extension programs in the United States, obstacles (ordered most critical to least critical) to agroforestry adoption include:

- Lack of developed markets.
- Unfamiliarity with technologies.
- Lack of awareness.
- Competition between trees, crops and animals.
- Lack of financial assistance.
- Lack of apparent profit potential.
- Lack of demonstration sites.
- Expense of additional management.
- Lack of training or expertise.
- Lack of knowledge about where to market products.
- Lack of technical assistance.
- Adoption/start up costs, including costs of time.
- Unfamiliarity with alternative marketing approaches(e.g. web).
- Unavailability of information about agroforestry.

- Apparent inconvenience.
- Lack of infrastructure(e.g. buildings, equipment).
- Lack of equipment.
- Insufficient land.
- Lack of seed/seedling sources.
- Lack of scientific research.

TREE CROP INTERACTION IN AGROFORESTRY SYSTEM

Interaction is defined as the effect of one component of a system on the performance of another component and the overall system. Regarding this, ICRAF researchers have developed an equation for quantifying treecrop interaction (I), considering positive effects of tree and crop yield through soil fertility enrichment (F) and negative effects through crop competition(C) for growth resources between tree and crop I=F-C. If F> C, interaction is positive, if F< C interaction is negative and if F=C interaction is neutral. Interaction occurs both above and below ground and includes a complex set of interaction relating to radiation exchange, the water balance, nutrient budget and cycling, shelter and other microclimatic modifications.

Interactions help to Know

- How the components of agroforestry utilize and share the resources of the environment.
- How the growth and development of any of the component will influence the others.

Factors affecting Tree Crop Interaction

- Effect of species: Proper choosing of compatible treecrop combinations.
- Effect of sun light: Light crown tree, either selection of shade tolerant crops or management of tree crop for reducing shade on agricultural crops.
- Effect of density: Numbers of trees/ha, planting of tree at optimum numbers of tree in a given area for reducing competition among crop and tree.
- Effect of age: At early stage of tree crop, competition is minimal.
- Effect of site factors: Relates about the carrying capacity of the site, site quality.

- Effect of management: Level of management for tree crop for benefits of agricultural crops or improving the total productivity of the system.

Advantages of Tree-crop Studies

- Choice of Species: Proper selection of both tree as well agricultural crops.
- Design of agroforestry system: Either parallel rows of trees and crops or concentric rows of crops around the tree.
- Management of agroforestry System: Degree of management, at what time, etc.

Negatives Effects

Competition

When plants grow in proximity to each other they interact either in positive ways (complementary) or in negative ways (competition). The biophysical bottom line of agroforestry is how to manage the interaction for light, water and nutrients between the tree component and the crop and livestock components for the benefit of the farmer. Competition may be above and below ground competition for resources uptake. However, the extent of below-ground competition is often not apparent.

Above Ground Competition

Competition for solar radiation is the most prominent above ground competition between trees and companion crops. Low light intensity is one of the important constraints for higher yield. Dhillion et al. concluded that the causes of reduction in growth and yield losses due to Eucalpytus tree plantation was due to direct competition for moisture, light and nutrients from the nearby rows of pear trees.

Below-ground Competition

Tree roots can compete with annual crop roots for available water and nutrients in the top soil. Below ground root competition for moisture, nutrients and space is relatively more important in agroforestry systems than above ground crown competition. Since light is more relatively more abundantly available than moisture and nutrients. It is necessary to have information on the nature of root development in two types of crop plants.

Dadhwal and Tomar found tree root pruning was very helpful in enhancing crop yields in marginal and degraded lands of north-western Himalaya of Uttaranchal thus indicating reducing in the root competition.

Allelopathy

The phenomenon of one plant having detrimental effect on another through the production and exertion of toxic chemical compounds is called 'allelopathy'. Allelopathy substance was first detected by Davis in black walnut whose foliar leachate containing Juglone was found to damage germination and seedling growth of crops beneath the tree. Allelopathy is one of the widely considered limitations for promotions and adoption of agroforestry at the field scale.

Complementary Effect (Positive Effect)

There are a several complementary effects of tree crop interaction such as increased productivity, improved soil fertility, efficient and balanced nutrient cycling, improved Soil conservation management and improvement of Microclimate which are very important in the way of overall agroforestry health and its productivity.

NITROGEN FIXATION THROUGH AGROFORESTRY

Agroforestry represents an approach to integrated land use involving the more or less intimate association of different plant species, always including trees or woody perennials on the same unit of land. This definition encompasses the association of trees with perennial crops, such as coffee or cacao, pasture species, annual or biennial crops, trees being planted as windbreaks or in alleys and frequently trimmed during cropping, various forms of multispecies combinations such as those found in homegardens, and also mixed forests (mixed perennial cropping).

The most decisive factor for the success of agroforestry is the choice of suitable, useable tree species. This choice should be based on economic and agronomic criteria. In order to fulfil the second criterion, the highest priority should be given to selecting trees that can improve the soil, and identifying species or clones that will fix or absorb large amounts of N (and also other elements, especially P) and then return them to the soil. Nitrogen-fixing tree species are probably the best choice, if they can actively fix nitrogen and thus significantly contribute to the improvement of the nitrogen status of the soil. Nitrogen-fixing trees can increase the yield of associated crops (annual or perennial) through different mechanisms:

- The first and main one, which is specific to nitrogen-fixing trees, consists of fixing atmospheric nitrogen and contributing nitrogen to the soil via leaf and fallen fruit litter (litter *sensu stricto*) or the release of root debris and nodules (root litter).
- The second one consists of the concentration of soil nutrients extracted from the deeper soil horizons, and eventually from the water table, and their return to the soil surface with the litter.

In addition, a series of indirect mechanisms (i) affect soil physical, chemical and biological properties, such as structure, or nitrification, whose reduced intensity alleviates the inhibiting effect of nitrates on nitrogen fixation; (ii) control diseases or pests, such as root nematodes (Mulongoy, personal communication); or (iii) stimulate the activity of beneficial organisms such as earthworms. In this chapter the discussion is restricted to the nitrogen-fixation process, starting with the most promising nitrogen-fixing species for the tropics, and continuing with current management techniques, and trends and directions for future agroforestry research.

Nitrogen-fixing Characteristics of the Main Tree Species used in Agroforestry

Legumes

Only nodulating, that is potentially nitrogen-fixing, legume trees are dealt with in this section. No attention is given to species such as *Parkia biglobosa* or *Cassia siamea* which do not nodulate, and consequently do not fix nitrogen, but are nevertheless, sometimes integrated in agroforestry systems.

Tropical legumes may nodulate with two types of *Rhizobium*: fast-growing strains which belong to the genus *Rhizobium* (*sensu stricto*), and slow-growing strains which form the cowpea miscellany and are now designated as *Bradyrhizobium*. One category of trees nodulates only with Rhizobium (*sensu stricto*), e.g., *Leucaena leucocephala* (with a few exceptions) or *Sesbania grandiflora*. Another category nodulates only with *Bradyrhizobium*, e.g., *Acacia meamsii* or *A. albida*. A third category is more promiscuous since it nodulates both with *Rhizobium* and *Bradyrhizobium*, e.g., *Acacia seyal*.

Acacia

Acacia albida (syn. *Faidherbia albida*) This tree, native to Africa, is usually considered as a highly valuable component of agroforestry systems, not only as a soil improver, but also as a source of fuelwood and forage in semi-arid zones, especially in Africa.

The yield of crops such as sorghum, peanut, or maize grown under a tree canopy of *A. albida* is, on the average, substantially higher than when grown alone, which indicates that the tree improves the soil fertility. However, it is not known whether the beneficial effect is due to nitrogen fixation or to other mechanisms enumerated in the introduction. *Acacia albida* nodulates with strains of *Bradyrhizobium*. Since these strains are already present in most soils, *A. albida* can be expected to respond poorly to inoculation except in sterilized nursery soils. Nodulation is normally observed on young seedlings, but nodules are seldom found in the field, which suggests that the nitrogen-fixing potential of this *Acacia* is rather low. However, this conclusion requires confirmation through precise measurements. The nitrogen-fixing potential of *A. albida* could possibly be improved by capitalizing on its great genetic variability.

Acacia Senegal This is another African Acacia species which is intercropped with food crops in the semi-arid zone. *A. Senegal* nodulates only with fast-growing strains of *Rhizobium*, i.e., *Rhizobium* (*sensu stricto*). Since these strains are less ubiquitous than strains of *Bradyrhizobium*, one can predict that *Acacia Senegal* will require inoculation more often than *A. albida*. The nitrogen-fixing potential of *A. Senegal* has not yet been estimated.

Acacia nilotica, A. raddiana (syn. *A. tortilis*) and *A. seyal* Like *A. albida* and *A. Senegal,* these acacias are often integrated into different types of agroforestry system in the semi-arid zones. *A. nilotica* and *A. raddiana* nodulate with fast-growing *Rhizobium* whereas *A. seyal* nodulates with both fast- (Rhizobium, sensu stricto) and slow-growing (*Bradyrhizobium*) strains. The nitrogen-fixing potential of the three species is still unknown.

Acacia auriculiformis and *A. mangium A. auriculiformis* has been reported to be used in agroforestry fuelwood production systems in Papua New Guinea. It produces profuse bundles of nodules, which suggests a good nitrogen-fixing potential.

Acacia mangium hybridizes naturally with its close relative *A. auriculiformis* . *A. mangium* is assumed to be a good nitrogen fixer and, hence, a promising component in fuelwood production systems when grown in appropriate climatic zones.

Acacia meamsii (syn. *A. mollissima* or *A. decurrens* var. *mollissima*) A highland tree native to Australia, *A. meamsii* has been introduced in many countries for its bark which is very rich in tannins. It is also frequently found as a multipurpose tree on farmlands in Java and eastern Africa. *A. meamsii* nodulates profusely with strains of *Bradyrhizobium* , even in very poor soils, provided the pH is not lower than 4.5. Nitrogen fixation is high: it was estimated to be approximately 200 kg N_2 ha_{-1} yr^{-1} by Orchard and Darby. A similar figure was given recently by Wiersum.

Albizia Lebbeck *and A.* Falcataria

There are about 100 species of *Albizia* distributed throughout Africa, Asia and tropical America, Two species, *Albizia lebbeck and A. falcataria,* are ideally adapted to agroforestry systems; they are renowned as soil improvers because of their profuse nodulation.

Native to India, Bangladesh, Burma and Pakistan, *Albizia lebbeck* is widely cultivated in tropical and subtropical regions with an annual rainfall as divergent as 500 and 2,000 mm. Perennial nodules of large size have been observed on adult trees in Senegal.

Native to the eastern islands of the Indonesian archipelago and the west of Irian, *Albizia falcataria* (syn. *Paraserianthes falcataria*) has been spread throughout South-East Asia. It is one of the fastest growing trees in the world provided that it is grown on sites where annual rainfall is high enough (2,000-2,500 mm). It nodulates abundantly, which suggests a good nitrogen-fixing capacity. However, because it is exacting in its soil requirements, *A. falcataria* is probably a poor nitrogen-fixer when it is introduced in relatively infertile soils.

Calliandra Calothyrsus

Native to Central America, this legume was introduced in Indonesia in 1936 with seeds from Guatemala, and has been shown to nodulate with a fast-growing strain of *Rhizobium*. Although prized as a first-class soil improver in rotation schedules and in intercropped systems, considerably more research is needed to accurately assess its nitrogen-fixing potential and its use as forage and organic fertilizer.

Erythrina spp.

More than 100 species are planted as shade trees, windbreaks, living fences, plant supports, for food and even medicinal purposes, or as components of alley-cropping systems. *E poeppigiana,* a fast-growing species, is widely used for shade in coffee and cacao plantations. It has been shown to nodulate with a strain of *Bradyrhizobium.* Nodules of *E poeppigiana* tend to be large, spherical, and clustered on the central root system. The biomass of the root nodules varied from 80 to 205 mg (dry weight) dm^{-3} soil, being highest close to the stem of the tree. A conservative estimate made in Venezuela and based on the decomposition of nodules during the dry season, indicated that the rate of nitrogen fixation was approximately 60 kg N_2 ha^{-1} yr^{-1}.

Gliricidia Sepium

Native to Central America, this species is widely used for shading cacao or coffee plantations. It has been shown to nodulate with fast-growing *Rhizobium*. Further experiments are needed to confirm this observation. Estimates of nitrogen fixation based on nodule biomass and rates of nitrogenase activity are approximately 13 kg N_2 ha^{-1} yr^{-1} in the conditions prevailing in Mexico. *G. sepium* has been introduced in western Africa in alley-cropping systems, but its nitrogen-fixing activity may be impeded by attacks of root nematodes. Sumberg reports that different accessions of *G. sepium* exhibit considerable variation. This large genetic variability should be exploited to improve nitrogen fixation.

Inga Jinicuil

This species, often found in the same sites as *Inga vera,* is a popular shade tree in coffee plantations in Mexico. In a plantation in Xalapa, annual nitrogen-fixation rates, based on the acetylene reduction method, were 35-40 kg N_2 ha^{-1} yr^{-1}, which, when compared to nitrogen from fertilizers, represents an important nitrogen input. The corresponding nodule biomass was 71 ± 14 kg (dry matter) ha^{-1}. Given a density of 205 trees ha^{-1}, nodule biomass per tree was 346 g (dry weight), a figure similar to that reported by Akkermans and Houvers for Alnus.

Alley cropping based on the use of Gliricidia septum at the IITA station.

Nodule of Inga jinicuil (a leguminous tree). The nodule results from the infection of the roots by the nitrogen-fixing bacterium, Rhizobium.

Mimosa Scabrella

Native to the Parana region of south-east Brazil, *Mimosa scabrella* is used in humid and subhumid tropical highlands as a multipurpose tree on farmlands and as a fuel-wood producer in agroforestry operations. It was shown to nodulate with a fast-growing strain of *Rhizobium. Mimosa scabrella* responds positively to inoculation, but its exact nitrogen-fixing potential has not yet been evaluated.

Leucaena Leucocephala

This tree has been the focus of a great deal of research in the past few decades. Native to Central America, it has been planted in many tropical countries, including south-east Asia, Africa and South America, as a shade tree for commercial crops, alley cropping or agroforestry wood production. The acetylene reduction method and the difference method, which have been used to evaluate nitrogen fixation by *Leucaena leucocephala,* give figures in the range of 100-500 kg N_2 ha^{-1} yr^{-1}. These figures have been confirmed recently by Sanginga, Mulungoy and Ayanaba (personal communication), who used the ^{15}N dilution method to make a precision evaluation of the nitrogen-fixation rate of *Leucaena leucocephala* grown in an Alfisol, pH 6.1, at the International Institute of Tropical Agriculture (IITA) in Ibadan, Nigeria. They showed that *Leucaena leucocephala* fixed 98-134 kg N_2 ha^{-1} in 6 months. The high nitrogen-fixing potential of this tree is related to its abundant nodulation under specific soil conditions, in which the nodule dry weight was reported to reach approximately 51 kg ha^{-1} in a stand of 830 trees ha^{-1}, and approximately 63 kg ha^{-1} in a stand of 2,500 trees ha^{-1} .

Leucaena leucocephala generally nodulates with Rhizobium (*sensu stricto*), and occasionally nodulates with *Bradyrhizobium*. The *Rhizobium* strain specific to *Leucaena leucocephala* is not generally found in soils. This explains the positive response to inoculation obtained in most soils where the level of nutrients (other than nitrogen) is high enough to satisfy the-tree's requirements.

Leucaena leucocephala is not a miracle tree. Its sensitivity to soil acidity and its high nutrient demand are reflected in its poor performance in infertile soils, e.g., the sandy soils of the Pointe Noire region of the Congo or of Hainan Island, China, even when properly inoculated and grown in a suitable climate.

POTENTIAL OF TREE AND SHRUB LEGUMES IN AGROFORESTRY SYSTEMS

Global agriculture is facing a series of challenges mainly related to growing population, climate changes and loss of biodiversity. Firstly, it is estimated that crop production must increase more than 60% by the year 2050 to fulfil the needs of the world's population. Secondly, drought and soil salinization are expected to result in losses of up to 50% of arable lands by the middle of this century. Thirdly, the spreading of agriculture to arid and semi-arid regions under intensive irrigation management will promote secondary soil salinization. Thus, the future of agriculture must rely on the sustainable intensification of crop production to feed the increasingly growing population, as well as on the use of tolerant cultivars that are able to cope with extreme environmental conditions, i.e. low fertility and saline soils, increasing water shortage periods as well as raising air temperatures and CO_2.

These challenges will be particularly critical in the developing countries, which have the highest rates of population growth and where most of the farmland is managed by smallholders. It is estimated that in these countries one of five persons still live on less than $1.25 a day. In this context, intensive agriculture based on agro-chemicals and mechanization is not sustainable and the systems must rely on appropriate cropping and post-harvest practices, preferably based on local ecosystem-based resources. Such practices include, for example, the implementation of integrated agroforestry systems, crop-livestock integration and crop-aquaculture production, that concomitantly have the potential to promote the conservation and the rational use of biodiversity and other ecosystem services.

According to the Food and Agriculture Organization of the United Nations (FAO), sustainable agriculture lies at the core of the 2030 Agenda. Indeed, 6 of the 17 sustainable development goals (SDGs) concentrate on this issue. These are as follows: (i) SDG 2—End hunger, achieve food security and improved nutrition and promote sustainable agriculture; (ii) SDG 6—Ensure sustainable consumption and production patterns; (iii) SDG 12—Ensure sustainable consumption and production patterns; (iv) SDG 13—Take urgent action to combat climate change and its impacts; (v) SDG 14—Take urgent action to combat climate change and its impacts; (vi) SDG 15—Sustainably manage forests, combat desertification, halt and reverse land degradation, halt biodiversity loss. Besides that, all the other 11 SDGs cross cut issues towards the end of hunger and poverty.

Since 2014, FAO has supported over 80 initiatives in Africa to promote sustainable agricultural production practices. To achieve that, three intertwined pillars are considered essential: (i) efficient use of resources, i.e. agriculture intensification to produce more with less impact on natural resources; (ii) environment protection and conservation, i.e. better management of natural resources in order to protect biodiversity (and ecosystem's stability), water, soil fertility and reduce pollution and (iii) resilient agriculture, i.e. adopting approaches to adapt and mitigate the impact of climate change.

Legume fixing trees and shrubs play a crucial role in biodiversity dynamics. From the ecological point of view, their introduction in cropping systems may contribute to reduce the use of chemical fertilizers and to ecosystems stability.

The Fabaceae or Leguminosae family is the third largest group of flowering plants and the second most important in agriculture. According to recent molecular and morphological studies, Fabaceae is a single monophyletic family, comprising more than 18,000 species distributed over ca. 800 genera and six sub-families: (i) Cercidoideae and (ii) Detarioideae, both comprising mainly tropical species; (iii) Duparquetioideae, a sub-family from western and central Africa, with only one species identified; (iv) Dialioideae, widespread throughout the tropics; (v) the pantropical Caesalpinioideae, with more than 4000 species (including the former sub-family Mimosoideae) and (vi) the cosmopolitan and largest legume sub-family, Faboideae (Papilionoideae), with ca. 14,000 species, mainly herbs and small shrubs.

Table: Sub-families, number of genera and species, distribution and examples of tree and shrub legumes.

Subfamily	Genera (number)	Species (number)	Distribution
Cercidoideae	12	ca. 335	Mainly tropical, e.g. Bauhinia spp., Cercis spp.
Detarioideae	84	ca. 760	Mainly tropical, e.g. Amherstia spp., Detarium spp., Tamarindus spp.
Duparquetioideae	1	1	West and Central Africa, Duparquetia orchidaceae
Dialioideae	17	ca. 85	Widespread throughout the tropics, e.g. Dialium spp.

Caesalpinioideae*	148	ca. 4400	Pantropical, e.g. Caesalpinia spp., Senna spp., Mimosa spp., Acacia spp.
Faboideae (Papilionoideae)	503	ca. 14,000	Cosmopolitan, e.g. Astragalus spp., Lupinus spp., Pisum spp.

Varying in habit from annual herbs to large trees, legumes are conspicuous and well represented throughout temperate and tropical regions. The family is particularly diverse in tropical forests and temperate shrub lands with a seasonally dry or arid climate. Such preference for semi-arid to arid habitats seems to be related to a nitrogen-demanding metabolism. The vast majority of legume species (ca. 90%) is able to establish symbiosis with nitrogen-fixing diazotrophic bacteria of the genera Rhizobium or Bradirhizobium (collectively called rhizobia) at the root and, in some cases, at the shoot level. The symbiosis results in the formation of a new plant organ, i.e. the root- or stem-nodule, where bacteria are hosted and fix atmospheric N_2, receiving in exchange energy and carbon to sustain their own metabolism as well as the symbiotic process. This type of symbiosis has around 58 million years and arose from the genome duplication of the sub-family Papilionoideae.

Details from Brachystegia boehmii leaves (A); Brachystegia spiciformis leaves and flowers (B); Cajanus cajan leaves, flowers and pods (C); Pterocarpus angolensis young leaves and mature fruit (D). Credits to Moura (A, D) and Catarino (B, C).

Nitrogen is among the key elements for plant growth and production, being decisive to the adequate plant response to environmental stresses. It is a major component of chlorophyll (photosynthesis), purines and pyrimidines (nucleic acids), amino acids (proteins) and ATP (energy). Although it is one of the most abundant elements in the Earth, its predominant form, i.e. N_2 (g), cannot be directly assimilated by the plants, which need reduced forms of this element (NH_4^+, NO_2^- and NO_3^-). This conversion can be achieved chemically through the Harber-Bosch process, or biologically through bacterial nitrogen fixation. While chemical nitrogen fixation is cost intensive and 40–50% of the nitrogen applied as fertilizer is lost via denitrification, runoff or leaching, only 10–20% of the biologically fixed nitrogen is lost that way. Besides that, the use of chemical fertilizers has a series of ecological impacts, such as air, soil and water pollution. Thus, there is a strong interest in symbiotic N_2 fixation between legumes and rhizobia towards the improvement of agricultural systems, i.e. better productivity with the least ecological impact.

In Africa, tree and shrub legumes provide a wealth of goods and services (e.g. wood, food, medicines, energy and housing) to millions of rural and urban dwellers. The interest on this group of legumes has increased tremendously in the last decades, particularly regarding soil erosion control and farming systems (e.g. windbreaks, shade trees, nitrogen fertilizers, forage, fruits and vegetables).

Table: Examples of tree and shrub legumes and their applications in formal and informal economies.

Species	Applications	Origin	$kg.N.ha^{-1}.yr^{-1}$
Acacia senegal (L.) Willd.	Poles, household, agriculture crafts, firewood, charcoal (stem and branches); tannin, ropes (bark); food (pods and seeds); forage (foliage and pods); honey (flowers); gum Arabic (Gum), medicine (various); erosion control; nitrogen fixation; fertilizer; fencing; intercropping	Drier tropical Africa, from Senegal and Mauritania (west) to Eritrea and Ethiopia (northeast) and South Africa (south); Oman, Pakistan and India	28.7–46.7 7–12 less than 20
Brachystegia boehmii Taub.	Small articles, firewood (stem and branches); ropes, twine, cloth and fishing nets, tanning, beehives (bark); food for edible larvae (leaves); medicinal (various)	Angola, Botswana, Malawi, Mozambique, Tanzania, Zaire, Zambia, Zimbabwe	Not available
Brachystegia spiciformis Benth.	Construction, furniture, household items, firewood, charcoal (stem and branches); tannin, beehives, ropes, sacks (bark); forage (foliage and pods); honey (flowers); medicinal (various); nitrogen fixation, shading	Angola, Kenya, Malawi, Mozambique, Tanzania, Zaire, Zambia, Zimbabwe	Not available
Cajanus cajan (L.) Millsp.	Light construction, baskets, fuel (stems and branches); forage (vegetative parts); honey (flowers); food (seeds and pods); medicinal (various); erosion control; shading, sheltering, nitrogen fixation; fertilizer; intercropping	Unknown origin, probably Indian and African; India	260 86 96 142
Gliricidia sepium (Jacq.) Walp.	Farm implements, furniture, posts, firewood, charcoal (stem and branches); forage (foliage and pods); honey and food (flowers); medicine and rodenticide (various); erosion control; shading; nitrogen fixation; fertilizer; fencing	Central America, Caribbean, South America, Asia (Java and Peninsular Malaysia)	210 35–38 108

Leucaena collinsii Britton & Rose1	Timber, firewood (stems and branches); forage (leaves); food (seeds); similar to gum Arabic (Gum); shading; nitrogen fixation; fertilizer; fencing; intercropping	Mexico and Guatemala	Not available
Pterocarpus angolensis DC.	Building, furniture and handicrafts (stem); fish poison (bark); body anointment (root bark); tannin, dyestuff (Sap); forage (foliage); honey (flowers); medicine (various); erosion control; nitrogen fixation	Angola, Mozambique, Namibia, South Africa, Swaziland, Tanzania, Zaire, Zambia	Not available
Sesbania sesban (L.) Merr.	Firewood, charcoal (Stem and branches); tannin, ropes (Bark); forage (leaves and young branches); food (flowers), ropes, fishnets (fiber); gum (seeds and bark); medicine (various); shading; fencing; fertilizer; nitrogen fixation; intercropping	Africa, Asia, Australia	84 100
Tephrosia candida (Roxb.) DC.	Firewood (stem and branches); forage and insecticide (leaves); erosion control, shading; land reclamation; nitrogen fixation; fertilizer; fencing; intercropping	India, SE Asia	Not available
Tephrosia vogelii Hook.f	Fish poison, insecticide and molluscicide (leaves); medicine (various); shading; fencing; nitrogen fixation; fertilizer	Tropical Africa, SE Asia	150

Importance and Role in Agroforestry Tropical Systems

Most tree and shrub legumes are resilient to extreme environments, e.g. erosion, low fertility, salinity, drought, fire and other adverse conditions. Such abilities seem to be innate and enhanced by the symbiosis with N_2-fixing rhizobia. According to Diabate et al. and Sprent, the use of nitrogen-fixing tree and shrubs (NFTSs) constitute a promising strategy to recover soil fertility, representing a sustainable agricultural approach to smallholder farmers. This is particularly important in sub-Saharan Africa where 80% of the farmland is managed by smallholders whose livelihoods depend strongly on the agricultural sector. Most of these households live below the poverty line and therefore cannot afford the use of fertilizers. For example, smallholders from Niger, Namibia and Mozambique use less than 1 $kg.N.ha^{-1}.yr^{-1}$, i.e., 100 times less than the average fertilizer needs for most crops.

The rates of N_2 fixation by NFTS depend on the species, climate and soil type, ranging from 0.1 to 700 $kg.N.ha^{-1}.yr^{-1}$. Despite the fact that many genera from the subfamilies Mimosoideae and Caesalpinioideae do not always establish root-nodule symbiosis, under proper environmental conditions, many species nodulate and fix atmospheric N at rates closer to those obtained with the traditional legumes belonging to the Papilionoideae. Additionally there is also evidence that NTFS are also able to increase P availability in the soil, mostly due to mycorrhizal associations.

The use of fertilizer tree legumes (Acacia anguistissima, Cajanus cajan, Gliricidia sepium, Leucaena collinsii, Sesbania sesban, Tephrosia candida and Tephrosia vogelii) for sustainable maize (Zea mays) production has been analysed by Akinnifesi and collaborators in East and Southern Africa (Zambia, Zimbabwe, Malawi and Tanzania). The authors reported a contribution of more than 60 $kg.N.ha^{-1}.yr^{-1}$ through biological nitrogen fixation (BNF), reducing the need of chemical N fertilizers in 75%. Besides that, N-fertilizer trees substantially increased crop yield, providing evidence that together with good management practices, maize yields can double as compared with traditional practices (without mineral fertilization). In Zambia, Mafongoya and Jiri have analysed the use of G. sepium as green manure for cabbage (Brassica oleracea) and onion (Allium cepa) production. This practice produced higher crop yields than the unfertilized and full rate fertilized controls: ca. 16 (unfertilized), 43 (full rate fertilized), 48 (gliricidia) and 65 $ton.ha^{-1}$ (half rate fertilizer and gliricidia) for cabbage and 22 (unfertilized), 43 (full rate fertilized), 65 (gliricidia) and 85 $ton.ha^{-1}$ (half rate fertilizer and gliricidia) for onion. In addition, gliricidia biomass replenished the soil with residual N, which was used by a subsequent crop, maize. In this case, the yields obtained (ca. 3–5 $ton.ha^{-1}$) were similar or slightly higher than those obtained with full rate fertilizer (ca. 2.5 to 4 $ton.ha^{-1}$) and unfertilized crop (ca. 1.5–3 $ton.ha^{-1}$). Nevertheless, caution should be paid to the potential environmental hazard of NO_3 leaching and resultant eutrophication, as well as to the choice of the best crops/NFTS combinations and cropping type (intercropping, multistrata or fallows).

Another interesting NFTS-based agroforestry system is the tree cropping system, like those used for coffee production. Such system is very popular in Latin America and less exploited in Africa. In Mexico, the rates of N_2 fixation obtained by Inga jinicuil in coffee plantations were above 40 $kg.N.ha^{-1}.yr^{-1}$, corresponding to 53% of the average amount of fertilizer applied annually. This observation reinforces the importance of the use of non-crop legumes in coffee agro-ecosystem. According to the literature, Inga spp. is the most popular choice from Mexico to Nicaragua. G. sepium and Erythrina poeppigiana are often the common choice in the low-lying areas of Honduras and Nicaragua and Costa Rica, respectively. In Africa, similar systems may constitute a promising and sustainable solution to improve coffee (or fruits) productivity in the region.

Agroforestry system with Albizia sp.,
coffee and maize in Gorongoza, Mozambique.

Towards Scientific Knowledge

Legume research has been mainly focused on annual grain crops. Instead, a limited amount of knowledge has been produced in perennials.

Stable Isotopes

The use of stable isotopes at natural abundance levels has brought a new dimension to our understanding of plant physiology and ecology. Analyses of the relative natural abundances of stable isotopes of carbon ($^{13}C/^{12}C$), oxygen ($^{18}O/^{16}O$), nitrogen ($^{15}N/^{14}N$) and deuterium (D/H) have been used across a wide range of scales, from cell to community and ecosystem level, contributing much to our understanding of the interactions between biosphere, pedosphere and atmosphere.

In general terms, processes such as diffusion and enzymatic incorporation favour the lighter isotope and lead to depletion of the heavier isotope as compared to source material. Natural abundance of ^{15}N can provide valuable information about N sources used by plants and fluxes of N in the ecosystems. There has been some debate on the interrelationship between nitrogen natural abundance ($^{15}N/^{14}N$) in soils and plants, and the use of a tracer or indicator of fractionation during N-uptake, assimilation and transport. Indeed, a variety of fractionations may occur during processes related to nitrogen transformation in soils and plants, which may complicate source-sink relationships. For example, nitrification discriminates against ^{15}N more than N mineralization, which makes NH_4^+ isotopically heavier than the organic N from which it is derived. Additionally, the ^{15}N of a particular compound may change and, together with the complexity of the N geochemical cycle, the use of ^{15}N should be carefully evaluated when applied to natural ecosystems.

However, there are substantial evidence that the natural abundance ratios for $^{15}N/^{14}N$ in soil and plants are useful integrators of the types and turnover rates of N cycling. These ratios can indicate whether a range of plants have access to the same N source. For instance, differences in leaf ^{15}N can indicate differences in rooting depth or root characteristics, such as mycorrhizal or N-fixing root associations. Also, nitrification and plant uptake properties (such as timing and type of uptake) can be determined by the leaf ^{15}N signatures. Robinson developed a mixing model to account for contrasting N sources which provided useful insights on the quantification of biological N fixation in tree legumes. Since N_2-fixing species typically have ^{15}N signatures close to the atmospheric value (0%), which strongly differ from the ^{15}N signature of non-fixing species, ^{15}N can be used as a sensitive tracer of N flow within an ecosystem. This approach was successfully used in the oligotrophic Portuguese primary dunes utilizing foliar ^{15}N of the non-leguminous native shrub Corema album. As the invasive Acacia longifolia and the native Stauracanthus spectabilis were the only legumes co-occurring with C. album, with no further sources of organic matter, this system represents an ideal model to quantify the impact of A. longifolia invasion. Similar to other ericoid ^{15}N mycorrhizal plants, C. album exhibited depleted foliar values without legume influence which, together with its high abundance in this system, may function as a good monitoring plant for legume influence.

Systems Biology

Systems biology is an emerging approach applied to biological scientific research that focuses on the complex interactions within biological systems, frequently associated with the environmental conditions. The best known example is the Human Genome Project which allowed major advances in human genetics and in the development of new medical therapies. Systems biology, commonly called 'Omics' is associated with high-throughput analysis of e.g. genomes (DNA, genomics), transcriptomes (RNA, transcriptomics), proteomes (proteins, proteomics), metabolomes (metabolites, metabolomics), lipidomes (lipids, lipidomics) and interactomes (interactions between molecules, interactomics) coupled with bioinformatics, which integrates computational, statistical and mathematical modelling. In plants, systems biology has been essentially focused on models, such as arabidopsis and annual crops (e.g. rice, wheat, tomato, soybean, maize, sorghum, chickpea or groundnut. Systems biology research in perennial plants is still restricted to a small group of trees, namely eucalyptus, poplar, abies and pine. Among others, such studies led to significant advances on the global knowledge of plant biology (development and functioning), genomics-assisted breeding towards the production of crops tolerant to extreme temperatures, salinity, drought, pests and diseases, or the discovery of new bio-compounds with application in agriculture, medicine and in a wide range of industries.

An integrated approach, combining eco-physiology and system's biology to understand the responses of two tree legumes (Brachystegia boehmii and Colophospermum mopane) to abiotic stresses, namely high temperatures, drought and low soil fertility. Preliminary results indicate that these plants have an innate ability to cope with extreme environments and that such capacity is linked to an enhanced water and mineral use efficiency, reinforcement of the photosynthetic machinery and the antioxidant system as well as an elevated osmoprotection state.

Agriculture has a primordial role to fight poverty and hunger and increase crop resilience to climate changes. The introduction of tree and shrub nitrogen-fixing trees into cropping systems is the most straightforward approach to reduce the use of chemical fertilizers, improving the soil ecosystem and the livelihoods of smallholder farmers in southern Africa. Additionally, agroforestry may improve ecosystem services such as, soil organic matter, biodiversity and N-retention. However, it is not devoid of environmental consequences, specifically N-leaching. Therefore, the implementation of agroforestry systems with NFTS should be preceded by experimental assays, in order to identify the factors promoting N-losses and design appropriate management strategies that synchronize legume-N availability with the crop demand.

References

- The-physical-environment-within-forests, knowledge, library, scitable -24202119: nature.com, Retrieved 27 January, 2020
- Wojtkowski, Paul Anthony (2002). Agroecological Perspectives in Agronomy, Forestry, and Agroforestry. Science Publishers. ISBN 978-1-57808-217-9
- Energy-government-and-defense-magazines, forest-resources, environment: encyclopedia.com, Retrieved 06 March, 2020
- Short-notes-on-the-functions-of-forests, knowledge-111022: shareyouressays.com, Retrieved 15 April, 2020
- Muschler, R. G. (2001-08-01). "Shade improves coffee quality in a sub-optimal coffee-zone of Costa Rica". Agroforestry Systems. 52 (3): 253. doi:10.1023/A:1011863426305. ISSN 0167-4366
- Forestry-definition-and-types-of-forestry-1584: agriinfo.in, Retrieved 08 May, 2020
- Importance-components-agroforestry: conserve-energy-future.com, Retrieved 16 June, 2020
- Abugre, S.; Asare, A.I.; Anaba, J.A. (2010). "Gender equity under the Modified Taungya System (MTS). A case of the Bechem Forest District of Ghana" (PDF). International Journal of Social Forestry. 3 (2): 134–150 (137). Archived from the original (PDF) on 19 August 2015
- The-potential-of-tree-and-shrub-legumes-in-agroforestry-systems, nitrogen-in-agriculture-updates: intechopen.com, Retrieved 30 August, 2020

Classification of Agroforestry Systems

CHAPTER 2

Agroforestry can be classified into a number of systems such as silviculture, agri-silviculture system, silvopasture, dehesa, agri-silve-pastoral system, forest gardening, etc. The topics elaborated in this chapter will help in gaining a better perspective of the classification of agroforestry systems.

Agroforestry systems can be classified according to the following sets of criteria:

- Structural basis: Considering the composition of the components, including spatial admixture of the woody component, vertical stratification of the component mix arid temporal arrangement of the different components.
- Functional basis: This is based on the major function or role of the different components of the system, mainly of the woody components (these can be product, e.g., production of food, fodder, fuelwood and so on or protective, e.g., windbreak, shelter-belts, soil conservation and so on).
- Socioeconomic basis: Considers the level of inputs of management (low input, high input) or intensity or scale of management and commercial goals (subsistence, commercial, intermediate).
- Ecological basis: Takes into account the environmental conditions on the assumption that certain types of systems can be more appropriate for certain ecological conditions. There may be a set of AF systems for arid and semi-arid lands etc.

Classification of Agroforestry System on Structural Basis

The structural of a system can be defined in terms of its components and the expected role or function of each. In this system the type of component and their arrangement are important. Hence, on the basis of structure, AF systems can be grouped into two categories:

- Nature of components.
- Arrangement of components.

Nature of Components: Based on the nature of components, AF systems can be classified into the following categories;

- Agrisilvicultural systems.
- Silvopastoral systems.
- Agrosilv opastoral systems.
- Other systems.

The following systems can be included:

- Apiculture with Trees: In this system various honey (nectar) producing tree species frequently visited by honeybees are planted on the boundary, mixed with an agricultural crop. The main purpose of this system is the production of honey.
- Aquaforestry: In this system various trees and shrubs preferred by fish are planted on the boundary and around fish-ponds. Tree leaves are used as forage for fish. The main or primary role of this system is fish production and bund stabilization around fish-ponds.
- Multipurpose Wood Lots: In this system special location-specific MPTS are grown mixed or separately planted for various purposes such as wood, fodder, soil protection, soil reclamation etc.

Arrangement of Components

The arrangement of components gives first priority to the plants even in AF systems involving animals. Their management according to a definite plan, say a rotational grazing scheme, gives precedence to the plants over the animals. Such plant arrangements in multispecies combinations involve the dimensions of space and time.

- Spatial Arrangement: Spatial arrangements of plants in an AF mixture may result in dense mixed stands (as in homegardens) or in sparse mix stands (as in most systems of trees in pastures). The species (or species mixtures) may be laid out in zones or strips of varying widths. There may be several forms of such zones, varying from microzonal arrangements (such as alternate rows) to macrozonal ones.
- Temporal Arrangement: Temporal arrangements of plants in AF may also take various forms. An extreme example is the conventional shifting cultivation cycles involving 2-4 years of cropping and more than 15 years of fallow cycle, when a selected woody species or mixtures of species may be planted. Similarly, some silvopastoral systems may involve grass leys in rotation with some species of grass remaining on the land for several years. These temporal arrangements of components in AF are termed coincident, concomitant, overlapping (relay cropping), separate and interpolated.

Functional Classification of Agroforestry Systems

Two fundamental attributes of all AF systems are productivity and sustainability. This clearly indicates that AF systems have two functions.

- Productive functions (producing one or more products): The various productive functions of AF systems are:
    - Food,
    - Fodder,
    - Fuelwood,
    - Other woods,
    - Other products.
- Protective functions (protecting and maintaining production systems): The protective functions of AF systems are:
    - Wind-break,
    - Shelter-belt,
    - Soil conservation,
    - Moisture conservation,
    - Soil improvement,
    - Shade (for crop, animal and man).

Socioeconomic Classification of Agroforestry Systems

Based on such socioeconomic criteria as scale of production and level of technology input and management, agroforestry systems have been grouped into three categories:

- Commercial,
- Intermediate and
- Subsistence systems.

Ecological Grouping of Agroforestry Systems

Based on the major agroecological zones, agroforestry systems are grouped into the following categories:

- Humid/sub-humid lowlands: This region is characterized by hot humid climate for all or most of the year and evergreen or semi-evergreen vegetation. The lowland humid and subhumid tropics (commonly referred to as the humid tropics) are by far the most important ecological region in terms of the total human population. It supports extent of area and diversity of agroforestry and other land-use systems. Because of climatic conditions that favour rapid growth of a large number of plant species, various types of agroforestry plant associations can be found in areas with a high human population, e.g., various forms of home gardens, Plantation of crops with combinations and multilayer tree gardens, in areas of low population density, trees on rangelands and pastures.

- Semi-arid/arid lands: This region is characterized by rainfalls confined to 9-21 days in July -Sept., 2-4 wet months, vapour pressure deficit ranging from 9 mb in January to 30 mb in April May, solar radiation incidence (400-500 cal/cm^2/day), high wind velocity (20 km/hour), high potential evapotranspiration (6 mm/day) and high mean aridity Index (70-74.8%).

Highlands: Variable rainfall, degraded and shallow lands at high altitude to deep rich soils in valleys and great climatic variations are the features of highlands. This area is a storehouse of great biological diversity. The Himalayan region is an excellent example of this type of area. Agroforestry has long encompassed many well-known land-use systems practices.

SILVICULTURE

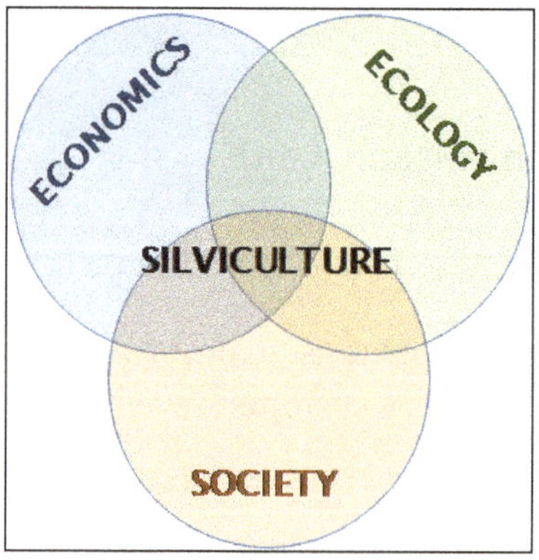

The three aspects of silvicultural decision-making.

Silviculture is the art and science of controlling the establishment, growth, composition, health, and quality of forests and woodlands to meet the diverse needs and values of landowners and society on a sustainable basis. It is often described as the tending of a forest or the growing of trees. Essentially silviculture is a discipline concerned with meeting human needs by manipulating a forest. Because forests typically require long time periods to grow, silviculture usually relies on management of ecosystems that closely mimic those found in nature. For this reason silviculture is often described as applied forest ecology. To understand silviculture you must understand the complex interactions of three key themes:

- Any management action must be ecologically feasible.

- Any management action must be conducted within all economic constraints.
- Any management action must consider the context of the governing society.

Historical examples of poor silvicultural decisions can invariably be attributed to the failure of the foresters involved to fully consider at least one of these critical themes. For instance:

- Litter raking, or removing leaves or needles from the forest floor, was a common management practice from the 12th until the early 20th century in Germany and central Europe. However, litter raking removed necessary nutrients from the forest floor and reduced subsequent growth of the trees. The foresters involved did not fully understand the effects of their actions on the ecology of these forests until the late 1800's, and thus managed in a way that was not sustainable on a long-term basis.
- The Monongahela National Forest clearcutting controversy in the 1960's and 1970's fundamentally changed how our National Forests are managed, and played a major role in the passage of the 1976 National Forest Management Act. While the foresters involved were making decisions that were both ecologically and economically sound, the controversy arose because they failed to fully consider the societal context in which they were operating. The result was public outcry, lawsuits, and more than a decade where forests went unmanaged while the controversy was settled.
- Cottonwood was widely planted across the US South from the 1960's to 1980's. Cottonwood is among the fastest growing hardwood species in this region, where plantation forests are a socially acceptable means of producing wood and fiber. However costs were eventually deemed prohibitive due to the intensive site preparation required. Relatively few plantations have been established since the 1980's.

It is always important to remember that forests are diverse and may be managed for many different goals:

- To maximize timber production and economic returns, as exemplified by loblolly pine plantations managed by corporations in the southern US.
- To create preferred wildlife habitat, as exemplified by the old-growth longleaf pine stands managed for red-cockaded woodpecker on many military bases in the southern US.
- To protect water quality, as exemplified by northern hardwood stands managed in the Catskill Mountains of New York to provide clean drinking water for New York City.
- To restore ecosystem function, as exemplified by bottomland hardwood stands replanted on former agricultural lands in the Lower Mississippi Alluvial Valley.

- To preserve natural forest ecosystems and wilderness, as exemplified by the management of forests in National Parks from the spruce-fir forests in Acadia, Maine to high elevation mixed stands in the Guadalupe Mountains in Texas, to coastal redwood stands in Redwood National Park in California.

Silvicultural systems may be applied to an individual stand to meet any one or several of these goals. While it is sometimes possible to meet more than one management objective on a single stand, it is not possible to optimize management for all possible objectives. Some objectives are inherently contradictory to one another (e.g. preservation and timber). The objectives of the forest landowner should always guide the forester in choosing the appropriate treatments for a stand. While not every stand is managed to maximize financial profits, every stand must be managed under whatever economic constraints are dictated by the landowner.

Depending on the management objectives, silvicultural inputs will vary in their intensity. For example, a late-successional northern hardwood stand in a state park being managed for recreation will likely require a low intensity of silvicultural inputs. The trees are growing relatively slowly due to the short growing season, and the stand structures desired are large, older trees that take a long time to grow. Thus rotation length will be long, there will likely be a high diversity of understory and midstory species, and stand structure (dead trees, canopy gaps, etc.) will be complex. Little may need to be done from a silvicultural standpoint, as this stand will naturally regenerate itself as new canopy gaps are formed in winter storms. However, there will be little profit from timber harvests.

Contrast this forest with a pine plantation in the US South being managed for timber. This stand will grow quickly due to the long growing season, will produce a large volume of wood, and thus will be harvested at a young age (i.e. a short rotation). To achieve this level of productivity, silvicultural inputs will be more intense, including site preparation, planting of genetically improved seedlings, fertilizer application, competition control with forest herbicides, and thinning. From an economic perspective, this stand will yield a greater rate of return and higher profits to the landowners. However, the stand will be very structurally simple and will have a lower diversity of understory and midstory vegetation, and perhaps less value as wildlife habitat.

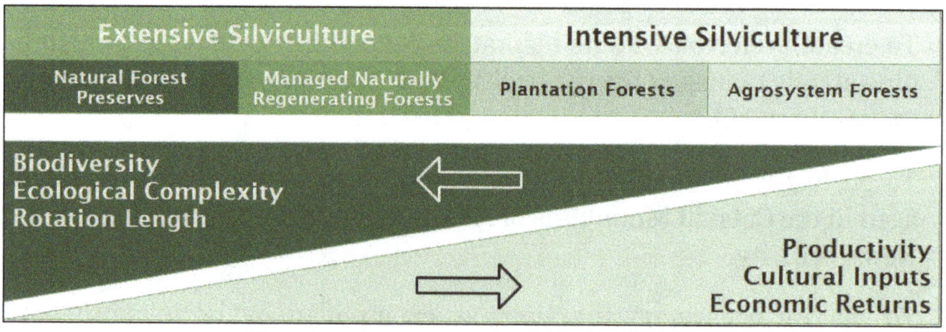

Silvicultural systems range from extensive to intensive in terms of their cultural inputs.

AGRI-SILVICULTURE SYSTEM

It is defined as growing of trees and agriculture crops together in same lands at the same time. This system is common in all agro-ecological zones of Nepal, where agriculture crops are grown in terrace flat and trees are grown in terrace bunds, borders and slopes. Trees grown in different regions are Terai (Sissoo, Eucalyptus, Baikaino, Ipil IPil etc.) hill (Mainly fodder trees such as Badahar, Tanki, Koiralo, Ipil-Ipil, Mulberry etc) and mountain (fodder trees such as Gogan, Dudhilo etc.).

The main advantages of this system are as follows:

- Produce multiple products such as food/vegetables/fruits, fodder and forage needed for livestock, fuel wood, timber, and leaf litter needed for organic manure production.
- Improve and sustain the crop productivity which increases the level of income of the farmers.
- Improve the nutritive value of animal feed due to the supply of green fodder.
- This is also the best practice for soil nutrient recycling, which also helps to reduce chemical fertilizer purchase.
- Improve the farm site ecology by reducing surface run off, soil erosion and nutrient loss, gully formation and landslides.
- Improve the local micro-climate and enhance the productive capacity of the farm.
- Reduce pressure of community forests and other natural forests for fodder, fuel wood and timber.
- This practices helps for the beautification of the surrounding areas.

Inter cropping practices can also adopt in this system in plain areas where trees should be grown maintaining rows to rows distance (5 meters) and plant to plant distance (2-3 meters). Management operations such singling, pruning and thinning in timber species and lopping, coppicing and pollarding in fodder species should be applied to reduce shading effects of trees on agriculture crops.

This system involves the conscious and deliberate use of land for the concurrent production of agricultural crops including tree, crops and forest crops. Based on the nature of the components this system can be grouped into various forms.

Improved Fallow Species in Shifting Cultivation

Shifting Cultivation

- It is prevalent in many parts of Africa, Latin America, South-East Asia and Indian subcontinent.

- It is known as 'jhuming' in North-east, 'khallu/kurwa' in Jharkhand.
- In this system, forest patch is selected and cleared felled. The herbs, shrubs and twigs and branches (slashed vegetation) are burnt. Cultivation of crops is done for a few years until soil fertility declines. The site is than abandoned (fallow period) and new patch is selected for cultivation of crops. The site is again cultivated after giving rest for few years.
- Earlier the fallow cycle was of 20–30 year. However, due to increasing requirement for cultivation of land due to population pressure, fallow period has reduced from 25–30 years to 2–3 years which has broken down the resilience of ecosystem and the land is increasingly deteriorating. Thus now shifting cultivation has become source of ecological degradation, soil erosion and converting good forests into wastelands.

Effect of Shifting Cultivation

- Deforestation and denudation of hill slopes-in secondary succession, area is occupied by weeds, useless shrubs etc.
- Soil erosion which leads to soil and nutrient losses, silting of reservoirs and streams, reduction in water-yield and landslips and landslides.
- Shifting cultivation adversely affects cation exchange capacity and physical properties of soil. It leads to lowering of organic matter and lowering the total quantity of sesquioxides, iron, aluminum, calcium, potassium, phosphorus, etc.
- Increases soil pH and reducing microbial activity.
- More weed growth and lower crop yield.
- No opportunity for infrastructural development.

Controlling Shifting Cultivation

- Motivate public for permanent agriculture by opening demonstration centers for improved agricultural practices, good quality seed, manuring, irrigation, weeding use of improved tools, terracing etc.
- Earning goodwill of local people: By engaging them in forest work and training them to undertake shifting cultivation on scientific lines.
- Arable land can be provided to the tribals for carrying out agriculture and also to settle in the area; a few schemes are being implemented under integrated tribal development programme.
- Legal measures: On steep slopes, near to roadside etc.
- Using land according to its capability.

- Provision of alternative management.
- Development of animal husbandry and dairy farming.
- Training of artisans and development of handicrafts.
- Employment in forest works and other industries.
- Providing communication facilities.
- Providing economic assistance for houses and agriculture operations.

Improved Fallow Species in Shifting Cultivation

- Fallows are crop land left without crops for periods ranging from one season to several years.
- The objective of improved fallow species in shifting cultivation is to recover depleted soil nutrients. Once the soil has recovered, crops are reintroduced for one or more season.
- The best species for the fallow system should induce good nitrogen fixation in the soil.
- The main aim of the fallow is to maintain or restore soil fertility and reduce erosion; some plants can be introduced primarily for their economic value.
- Plants included in improved fallows should be compatible with future crops, free of any negative physical or chemical effects on the soil and not in competition with the crops to be planted later on the same site.

FOREST GARDENING

Forest gardening is a low-maintenance, sustainable, plant-based food production and agroforestry system based on woodland ecosystems, incorporating fruit and nut trees, shrubs, herbs, vines and perennial vegetables which have yields directly useful to humans. Making use of companion planting, these can be intermixed to grow in a succession of layers to build a woodland habitat.

Forest gardening is a prehistoric method of securing food in tropical areas. In the 1980s, Robert Hart coined the term "forest gardening" after adapting the principles and applying them to temperate climates.

Forest gardens are probably the world's oldest form of land use and most resilient agroecosystem. They originated in prehistoric times along jungle-clad river banks and in the wet foothills of monsoon regions. In the gradual process of families improving their immediate environment, useful tree and vine species were identified, protected and improved whilst undesirable species were eliminated. Eventually superior foreign species were selected and incorporated into the gardens.

Robert Hart's forest garden in Shropshire.

Forest gardens are still common in the tropics and known by various names such as: *home gardens* in Kerala in South India, Nepal, Zambia, Zimbabwe and Tanzania; *Kandyan forest gardens* in Sri Lanka; *huertos familiares*, the "family orchards" of Mexico. These are also called agroforests and, where the wood components are short-statured, the term shrub garden is employed. Forest gardens have been shown to be a significant source of income and food security for local populations.

Robert Hart adapted forest gardening for the United Kingdom's temperate climate during the 1980s. His theories were later developed by Martin Crawford from the Agroforestry Research Trust and various permaculturalists such as Graham Bell, Patrick Whitefield, Dave Jacke and Geoff Lawton.

Temperate Climates

Robert Hart, forest gardening pioneer.

Hart began farming at Wenlock Edge in Shropshire with the intention of providing a healthy and therapeutic environment for himself and his brother Lacon. Starting as relatively conventional smallholders, Hart soon discovered that maintaining large annual vegetable beds, rearing livestock and taking care of an orchard were tasks beyond their strength. However, a small bed of perennial vegetables and herbs he planted was looking after itself with little intervention.

Following Hart's adoption of a raw vegan diet for health and personal reasons, he replaced his farm animals with plants. The three main products from a forest garden are fruit, nuts and green leafy vegetables. He created a model forest garden from a 0.12 acre (500 m²) orchard on his farm and intended naming his gardening method *ecological horticulture* or *ecocultivation*. Hart later dropped these terms once he became aware that *agroforestry* and *forest gardens* were already being used to describe similar systems in other parts of the world. He was inspired by the forest farming methods of Toyohiko Kagawa and James Sholto Douglas.

Seven-layer System

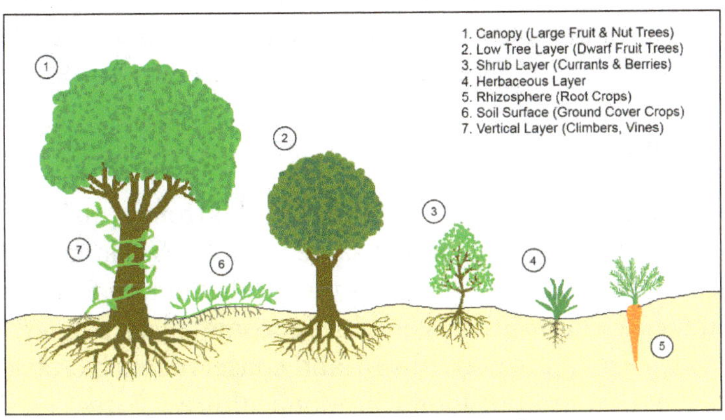

The seven layers of the forest garden.

Robert Hart pioneered a system based on the observation that the natural forest can be divided into distinct levels. He used intercropping to develop an existing small orchard of apples and pears into an edible polyculture landscape consisting of the following layers:

- 'Canopy layer' consisting of the original mature fruit trees.
- 'Low-tree layer' of smaller nut and fruit trees on dwarfing root stocks.
- 'Shrub layer' of fruit bushes such as currants and berries.
- 'Herbaceous layer' of perennial vegetables and herbs.
- 'Rhizosphere' or 'underground' dimension of plants grown for their roots and tubers.
- 'Ground cover layer' of edible plants that spread horizontally.
- 'Vertical layer' of vines and climbers.

A key component of the seven-layer system was the plants he selected. Most of the traditional vegetable crops grown today, such as carrots, are sun-loving plants not well selected for the more shady forest garden system. Hart favoured shade-tolerant perennial vegetables.

Permaculture

Bill Mollison, who coined the term *permaculture*, visited Robert Hart at his forest garden in Wenlock Edge in October 1990. Hart's seven-layer system has since been adopted as a common permaculture design element.

Numerous permaculturalists are proponents of forest gardens, or food forests, such as Graham Bell, Patrick Whitefield, Dave Jacke, Eric Toensmeier and Geoff Lawton. Bell started building his forest garden in 1991 and wrote the book *The Permaculture Garden* in 1995, Whitefield wrote the book *How to Make a Forest Garden* in 2002, Jacke and Toensmeier co-authored the two volume book set *Edible Forest Gardens* in 2005, and Lawton presented the film *Establishing a Food Forest* in 2008.

Tropical Climates

Forest gardens, or home gardens, are common in the tropics, using intercropping to cultivate trees, crops, and livestock on the same land. In Kerala in south India as well as in northeastern India, the home garden is the most common form of land use and is also found in Indonesia. One example combines coconut, black pepper, cocoa and pineapple. These gardens exemplify polyculture, and conserve much crop genetic diversity and heirloom plants that are not found in monocultures. Forest gardens have been loosely compared to the religious concept of the Garden of Eden.

The BBC's *Unnatural Histories* claimed that the Amazon rainforest, rather than being a pristine wilderness, has been shaped by humans for at least 11,000 years through practices such as forest gardening and *terra preta*. Since the 1970s, numerous geoglyphs have been discovered on deforested land in the Amazon rainforest, furthering the evidence of Pre-Columbian civilizations.

On the Yucatán Peninsula, much of the Maya food supply was grown in "orchard gardens", known as *pet kot*. The system takes its name from the low wall of stones (*pet* meaning 'circular' and *kot*, 'wall of loose stones') that characteristically surrounds the gardens.

In many African countries, for example Zambia, Zimbabwe, Ethiopia and Tanzania, gardens are widespread in rural, periurban, and urban areas and they play an essential role in establishing food security. Most well known are the Chaga or Chagga gardens on the slopes of Mt. Kilimanjaro in Tanzania. These are an example of an agroforestry system. In many countries, women are the main actors in home gardening and food is mainly produced for subsistence. In North Africa, oasis-layered gardening with palm trees, fruit trees, and vegetables is a traditional type of forest garden.

Projects

El Pilar on the Belize–Guatemala border features a forest garden to demonstrate traditional Maya agricultural practices. A further one acre model forest garden, called Känan K'aax (meaning 'well-tended garden' in Mayan), is funded by the National Geographic Society and developed at Santa Familia Primary School in Cayo.

In the United States, the largest known food forest on public land is believed to be the seven acre Beacon Food Forest in Seattle, Washington. Other forest garden projects include those at the central Rocky Mountain Permaculture Institute in Basalt, Colorado and Montview Neighborhood farm in Northampton, Massachusetts. The Boston Food Forest Coalition promotes local forest gardens.

In Canada Richard Walker has been developing and maintaining food forests in British Columbia for over 30 years. He developed a three acre food forest that at maturity provided raw materials for a plant nursery and herbal business as well as food for his family. The Living Centre has developed various forest garden projects in Ontario.

In the United Kingdom, other than those run by the Agroforestry Research Trust (ART), there are numerous forest garden projects such as the Bangor Forest Garden in Gwynedd, northwest Wales. Martin Crawford from ART administers the Forest Garden Network, an informal network of people and organisations who are cultivating forest gardens.

SILVOPASTURE

Silvopasture integrates livestock, forage, and trees.

Silvopasture is the practice of integrating trees, forage, and the grazing of domesticated animals in a mutually beneficial way. It utilizes the principles of managed grazing, and it is one of several distinct forms of agroforestry.

Properly-managed silvopasture can increase overall productivity and long-term income due to the simultaneous production of tree crops, forage, and livestock, and can provide environmental benefits such as carbon sequestration. Silvopasture is one of the oldest known forms of agriculture, and has been practiced in many parts of the world for centuries. Silvopasture is not the same as unmanaged grazing in woodlands, which has many known negative environmental consequences.

Benefits

A primary benefit of silvopasture is increased farmland utilization - silvopasture can incorporate unused woodlands into production and produce multiple products on the same acreage. This diversifies farm income sources and increases farm viability. Silvopasture has been found to increase wildlife abundance and diversity and to contribute to carbon sequestration and climate change mitigation.

Livestock

Trees in silvopasture systems provide livestock with protection from sun and wind, which can increase animal comfort and improve production. Trees can provide shade in the summer and windbreaks in the winter, allowing livestock to moderate their own temperature.

Certain tree types can also serve as fodder for livestock. Trees may produce fruit or nuts that can be eaten by livestock while still on the tree or after they have fallen. The leaves of trees may serve as forage as well, and silvopasture managers can utilize trees as forage by felling the tree so that it can be eaten by livestock, or by using coppicing or pollarding to encourage leaf growth where it is accessible to livestock.

Forage

Well-managed silvopasture systems can produce as much forage as open-pasture systems under favorable circumstances. Silvopasture systems have also been observed to produce forage of higher nutritive quality than non-silvopasture forage under certain conditions. Increased forage availability has been observed in silvopasture systems compared to open-pasture systems under drought conditions, where the combination of shade from trees and water uptake from tree roots may reduce drought impacts.

Trees

Silvopasture is compatible with fruit, nut, and timber production. Grazing can serve as a cost-effective vegetation and weed control method. Silvopasture can also help reduce pests and disease in orchards - when introduced into an orchard after harvest, livestock are able to consume unharvested fruits, preventing pests and diseases from spreading via these unharvested fruits and in some cases consuming the pests themselves.

Methods

Establishment

Silvopasture can be established by planting trees into existing pasture or by establishing pasture in existing woodland. These two establishment methods differ significantly.

Integrating Trees into Pasture

Planting trees into existing pasture presents several challenges: young trees must be protected from livestock, trees may take years to become productive (depending on the species), and planting trees in a pasture can limit the ability to use that land for other purposes in the future.

Integrating Pasture into Woodland

Integrating pasture into existing woodland presents challenges as well: the woodland likely needs to be thinned to increase light infiltration, which is time consuming and may require heavy machinery, as well as a strategy for dealing with felled trees. Thinned woodlands are also likely to experience a flush of growth in weeds and seedling trees that must be dealt with to prevent the pasture from being overgrown. Pasture forages may also need to be established beneath the trees, a process which can be difficult if trees have already been felled.

Wood pasture in winter in the Wisentgehege Springe game park near Springe, Hanover, Germany.

Ireland, Wales, Scotland and Lower

Tree species and planting densities are studied over a range of sites at The Silvopastoral National Network Experiment. Natural England's Environmental Stewardship scheme defines Wood Pasture, in the Farm Environmental Plan booklet, as a structure of open grown or high-forest in a matrix of grazed-grassland, heathland, and/or woodland floras.

Veteran pollard oak, a sign of ancient wood pasture at Windsor.

Their experience shows sheep use the trees for shelter from wind. This could provide significant animal-welfare benefits. However, 'sheep time' close to trees results in soil compaction with the greatest-compaction after trees are planted at very-low densities. Some botanists recommend trees be planted at no-less than 400 per hectare to ensure good-establishment.

Evidence of old wood-pasture management is detectable in many of the ancient woodlands of Scotland, such as Rassal Ashwood in Ross-shire, and at Glen Finglas in the Trossachs. The Dalkeith Old Wood, belonging to the Duke of Buccleuch, cattle grazing beneath ancient oak, is designated as a Site of Special Scientific Interest (SSSI) (or 'ASSI').

United States

Silvopastures are the most-viable and prominent agroforestry practice in the United States.

Silvopasture over the years.

DEHESA

A *dehesa* is a multifunctional, agrosylvopastoral system (a type of agroforestry) and cultural landscape of southern and central Spain and southern Portugal; in Portugal, it is known as a *montado*, Dehesas may be private or communal property (usually belonging to the municipality). Used primarily for grazing, they produce a variety of products, including non-timber forest products such as wild game, mushrooms, honey, cork, and firewood. They are also used to raise the Spanish fighting bull and the Iberian pig. The main tree component is oaks, usually holm (*Quercus ilex*) and cork (*Quercus suber*). Other oaks, including melojo (*Quercus pyrenaica*) and quejigo (*Quercus faginea*), may be used to form dehesa, the species depending on geographical location and elevation. Dehesa is an anthropogenic system that provides not only a variety of foods, but also wildlife habitat for endangered species such as the Spanish imperial eagle.

A dehesa in Bollullos Par del Condado, Huelva, southern Spain.

By extension, the term can also be used for this style of rangeland management on estates.

Nature

Dehesa in Extremadura, Spain.

The dehesa is derived from the Mediterranean forest ecosystem, consisting of grassland featuring herbaceous species, used for grazing cattle, goats, and sheep, and tree species belonging to the genus *Quercus* (oak), such as the holm oak (*Quercus ilex* sp. *ballota*), although other tree species such as beech and pine trees may also be present. Oaks are protected and pruned to produce acorns, which the famous black Iberian pigs feed on in the fall during the *montanera*. Ham produced from Iberian pigs fattened with acorns and air-dried at high elevations is known as *Jamón ibérico* ("presunto ibérico", or "pata negra" in Portuguese), and sells for premium prices, especially if only acorns have been used for fattening.

In a typical dehesa, oaks are managed to persist for about 250 years. If cork oaks are present, the cork is harvested about every 9 to 12 years, depending on the productivity of the site. The understory is usually cleared every 7 to 10 years to prevent the takeover of the woodland by shrubs of the rock rose family, often referred to as "*jara*", or by oak sprouts. Oaks are spaced to maximize overall productivity by balancing light for the grasses in the understory, water use in the soils, and acorn production for pigs and game.

There is debate about the origins and maintenance of the dehesa, and whether or not the oaks can reproduce adequately under the grazing densities now prevailing.

Importance and Economic Context

A dehesa in the Montes de Toledo.

The dehesa system has great economic and social importance on the Iberian Peninsula because of both the large amount of land involved and its importance in maintaining rural population levels. The major source of income for dehesa owners is usually cork, a sustainable product that supports this ancient production system and old growth oaks. High end black iberian pigs and sale of hunting rights also represent significant income sources. Periodic hunts in the dehesa are known as the *monteria*. Groups attend a hunt at a private estate and wait at hunting spots for game to be driven to them with dogs. They usually pay well for the privilege, hunting wild boar, red deer and other species.

The area of dehesa usually coincides with areas that could be termed "marginal" because of both their limited agricultural potential (due to the poor quality of the soil) and a lack of local industry, which results in isolated agro-industries and very low capitalization.

AGRI-SILVI-PASTORAL SYSTEM

Growing of trees, agriculture crops and grasses together in same lands at the same time is known as agri-silvi-pastoral system. This is typical hill farming system of Nepal, in which agriculture crops are grown in terrace flat, trees in terrace bunds, border and slopes and grasses in terrace slopes, and some farmers have been growing improved legume and non-legume grasses such as Mott napier (Pennisetum purpureum), Setaria (Seteria splendida), Mulato (Brachiaria brizantha x B. ruziziensis) and Forage peanut (Arachis pinotoi) along terrace bunds and borders.

The main advantages of this system are as follows:

- Produce multiple products such as food/vegetables/fruits, fodder and forage/ grasses needed for livestock, fuel wood, timber, and leaf litter needed for organic manure production.
- Improve and sustain the crop productivity which increases the level of income of the farmers.
- Improve the nutritive value of animal feed due to the supply of green fodder.
- This is also the best practice for soil conservation and soil nutrient recycling, which also helps to reduce chemical fertilizer purchase.
- Improve the farm site ecology by reducing surface run off, soil erosion and nutrient loss, gully formation and landslides.
- Improve the local micro-climate and enhance the productive capacity of the farm.
- Reduce pressure of community forests and other natural forests for fodder, forage/grasses, fuel wood and timber.
- This practices helps for the beautification of the surrounding areas.

References

- Wolfe, Martin S.; Pearce, Bruce D.; Smith, Jo (December 2012). "A European perspective for developing modern multifunctional agroforestry systems for sustainable intensification". Renewable Agriculture and Food Systems. 27 (4): 323–332. doi:10.1017/S1742170511000597. ISSN 1742-1713
- Agri-silvi-pastoral-system: forestryarticles.blogspot.com, Retrieved 10 August, 2020
- Whitefield, P. 2002. How to Make a Forest Garden. Hampshire: Permanent Publications. ISBN 1-85623-008-2
- Wilson, Matthew; Lovell, Sarah (2016-06-18). "Agroforestry—The Next Step in Sustainable and Resilient Agriculture". Sustainability. 8 (6): 574. doi:10.3390/su8060574. ISSN 2071-1050
- Crawford, Martin 2010. Creating a Forest Garden: Working with Nature to Grow Edible Crops. Totnes: Green Books. ISBN 1-900322-62-5

Silviculture: An Integrated Study

CHAPTER 3

The art, science and practice of controlling the structure, growth and quality of forests for meeting the needs of timber production is termed as silviculture. Coppice method, clearcut method, ecological thinning, selection cutting, seed tree method, etc. fall under its domain. This chapter has been carefully written to provide an easy understanding of the related concepts of silviculture.

SILVICS

Silvics is the term used for the characteristics that define the life history, growth, behavior and ecology of a tree species. It is often linked with silviculture, which is the application of silvics to the management of trees in order to enhance the reproduction, survival or growth of a specific tree species. Silvics describes all aspects of a tree species life cycle, from germination to death, including the environmental and soil conditions that are suitable for its growth, and the insects and pathogens that can harm it.

Forest biologists have spent many years developing an understanding of the characteristics of tree species for the simple reason that each tree species has a unique set of requirements for growth and development, as well as some unique growth attributes. Knowledge of these is essential for effective management of the species.

Management should be considered broadly in this context. Management may include intensive practices such as growing a species in short rotation monoculture plantations for fiber, but it can also mean any activity that enhances the growth and development of a species, for example, the establishment of a tree species in a new rural or urban setting, the reintroduction of a species where it once grew, improving the growth and vigor of specific tree species for the use of wildlife, the introduction of exotic species that will be of benefit, or the eradication of introduced species that have become noxious weeds.

An understanding of silvics also allows us to see the similarity and differences among tree species around the world. Trees can be classified into three general categories based on whether they grow in boreal, temperature or tropical regions. These classifications tell us that there is some similarity among species growing in the same climatic region. Some species are more able to grow and survive where it is cold (boreal), or where it remains warm all year long (tropical), or in that in-between zone where cold and warm alternate (temperate). Within each of these three climatic zones there are many, many tree species. Importantly, the silvics of each and very tree species is unique. Why?

The answer is that no matter how much we humans like to classify and lump things together into groups, tree species are distributed individually across the landscape, and are not distributed in packets of like species. That means that one species may grow best where it is wet, another were it is dry, another may have very specific soil nutrient or soil density requirements, while others grow well on many different types of soil. But each species has unique attributes and conditions allowing it survive and grow. These attributes are often called a species niche, which is an ecological term meaning the species' unique habitat or functional position in an ecosystem. While species can share some characteristics in common, and others may share many similar characteristics, no two species will have identical life histories, growth, behavior or ecological characteristics, i.e., no two tree species have the same silvics. Every tree species has some unique characteristics that make it successful in its environment.

The silvical characteristics of a tree species include life cycle, the environment and soil conditions in which it grows, and its response to biotic interactions such as competition or damaging insects. In the Silvics of North America edited by R. M. Burns and B. H. Honkala, these factors are divided into the categories of Habitat, Life History, Special Uses, and Genetics. Within these categories there are many factors that can differ among species and need to be considered.

Habitat

The habitat of a species is where it is found in nature. Habitat includes the species native range, or the historical geographical distribution of a species. It also includes the climatic range in which the species exists. Climate is determined by the pattern and total amount of precipitation, including whether it occurs as rain, snow or fog, or a combination of those forms of precipitation; and by temperature, characterized by average annual and growing season temperature and the maximum and minimum temperatures the species can tolerate; length of the growing season, or frost-free period; the type or types of soil the species is found on.

The total amount of solar radiation that a flat portion of the earth receives varies little on an annual basis, but it varies to a large extent seasonally and monthly. The response to patterns of solar radiation can help define a species, because the timing of dormancy, or the winter resting period, is linked to changes in day length. Whether a species can grow in the low light conditions of a forest understory or the high light conditions of an open disturbed site is a particularly important habitat characteristic.

Soils often exhibit a great variety of differences over small geographic areas, such as the depth of the soil, clay, silt and sand content, porosity, water holding capacity, and nutrient supply capacity. These differences can affect the suitability of the site for a tree species. Topography needs to be considered as well. Species often change from the bottom to the top of hills and mountains, and between sides of mountains. The causes of this change can be attributed to changes in the environment (precipitation, temperature, solar radiation) and soils across gradients in elevation or aspect. However, because topography integrates these factors together, it is often an excellent factor to use to help characterize a species. Species may occur singly or in association with a large number of other tree species. In colder, dryer or extremely wet sites fewer tree species are found. On moist or fertile sites, or sites with long growing seasons, often a large number of tree species are found growing together in the same forest. Tree species may also commonly be found with associated understory shrubs and herbs or grasses. The present, absence and number of tree and understory associates further define the unique characteristics of a tree species.

Species differences in life history begin with their seeds. Forest trees have a wide variety of seed characteristics. Seeds can be large or small, heavy or light, covered with a thick or thin outside protective coating or shell, dispersed by wind, water or animals, and range in dormancy release requirements from almost none to very specific heat, cold, moisture, aeration or light exposure before germination is achieved. Some conifers even require the heat of a fire to free the seeds from cones before dormancy release can be initiated.

Many of these dormancy factors are integrated into the concept of seed bed requirements of a species.

Favorable seed beds range from open disturbed areas, for example, after a fire, to the moist humus of the soil in an undisturbed forest. Seed beds across this range have different moisture, light, nutrient and temperature conditions, with each set of conditions favorable to different species. After germination, seedling development depends on the species ability to grow in the particular environment of the germination site.

Although thousands of seeds often germinate on a site, only a handful survive through the first or second years. This is often a time in which the seedling is exposed to intense abiotic and biotic stresses, and few seedlings are able to successfully establish in that situation. Understanding the characteristics that allow a particular species to flourish after germination is often the key to successful establishment. In intensive management situations, foresters avoid this critical period by germinating seeds in nurseries and greenhouses that offer a protected and resource rich environment to insure good survival. The seedling may be out-planted when it is one, two years old or older. This, coupled with proper site preparation to make the site favorable to the growth of the particular species insures a much higher rate of survival and faster initial growth rates.

Species are also characterized by their growth form. Aboveground, trees can be separated by whether they have evergreen or deciduous foliage. Evergreen species can retain foliage from less than two years to more than 45 years. In contrast, deciduous trees grow and shed their leaves every year. Both evergreen and deciduous species can have multiple flushes of foliage and branch growth during a single growing season, although some have only a single flush each year and others have continuous growth during the growing season. As trees age they can have fewer flushes than when they are young.

The height and age of tree species also is an important attribute. Maximum tree heights and longevities are species dependent and also vary with climatic region. For example, evergreen conifers in the Pacific Northwest of North America achieve much greater heights and longevities than evergreen conifers in eastern North America, Europe or Asia. Typically, trees that grow in more stressful environments are shorter, but they can have either deciduous or evergreen foliage displays. Tree species that have fast growth rates tend to have short lives and smaller stature, but as with all silvical characteristics, there are notable exceptions to these generalizations among tree species. The branching patterns of species also tend to be distinct, ranging from thick or thin, long or short, horizontal or upright, extending along the entire stem or occurring mostly at the top of the tree.

Bark color, thickness and pattern are also distinctive among species. Bark features help to protect the tree from insects and pathogens. Thick bark is a notable feature in trees that are resistant to fire. The growth rate of trees tends to indicate its successional status. Fast growth rates are advantageous in recently disturbed, early successional sites, whereas slow grow rates are commonly found in later successional species. For this reason most of the trees planted for wood or fiber production are in the early successional category.

Within the stem, wood exhibits additional important differences among species. Ring-porous wood consists of vessels that are evenly ranked from large to small each year. Diffuse-porous wood has vessels of different sizes evenly distributed throughout the annual ring. Tracheids, which are small diameter xylem cells, are yet another form of wood, mostly found in gymnosperms. Tree species with trachieds are called softwoods. Tree species with ring- or diffuse-porous wood are called hardwoods. However, the distinction is really not so clear. Some softwood species have more dense wood than some hardwood species. Species also have resin ducts in the wood for protection from insect attack. Wood can vary greatly in its density and flexibility, affecting its suitability for use as building materials, furniture or paper manufacturing. This makes wood traits particularly important in production forestry species.

Species can also be characterized by their patterns of root growth. Some species have large tap roots, others none at all. Some species are shallowly rooted, other deeply rooted. Some species develop large and extensive root systems; others have much smaller root systems. Even the size and color of the roots of different species can differ. In some instances tree species have very distinct and easily observed roots, such as the large buttress roots in mangroves or the "knee-roots" of bald cypress trees.

Reaction to biotic factors is also an important difference among species. Competition occurs between plants when the resources needed for growth are in insufficient supply for all plants to achieve their maximum growth rate. Tree species can be ranked by their ability to survive in various levels and types of competition. The competitive ability of a species is difficult to characterize, but extremely important. In the absence of competition, many tree species can survive and grow in a much wider range of conditions than those in which they are actually found in nature. Species have evolved in association with different insects and diseases. In some instances these damaging agents, such as root and stem rots, boring insects, or leaf herbivores prevent the species from achieving its maximum growth rate or potential longevity.

Special Uses

It is often the ability to use a tree species for a specific purpose that makes its silvical characteristics of interest. It might be for an ecologically oriented purpose such as habitat restoration, creation of forested wetlands, seed production for wildlife, or plantings for soil stabilization to prevent erosion. Some species can agricultural crops from wind damage, and some are useful for surface mine site reclamation. Others make excellent firewood for heating, cooking or charcoal production. Yet other species are valued for their wood properties for making different types of paper, furniture, musical instruments, stencils, chemicals, or building materials for housing, boat and airplane construction. Yet others are valued for their beauty in residential plantings (or for their resistance to air pollutants). The list of applications of trees species for special and sometimes unique purposes is almost endless.

Genetics

There is genetic variation in many of the characteristics that make each species unique. Species with wide geographical and environmental ranges, and those whose ranges are not continuous, tend to have more genetic differences. In some instances they are so distinct as to be considered separate varieties. Genetic differences can be expressed in any traits, from time of budburst to timing of dormancy, wood properties, leaf, stem and root physiology, leaf shape, tolerance to environmental stress, susceptibility to insect predation or disease, growth rate and longevity. However, overall genotypes within a species share many more similarities than differences, so that they are still recognizable as members of the same species.

Silviculture

Silviculture is the practice of controlling the growth, composition/structure, and quality of forests to meet values and needs, specifically timber production.

The study of forests and woods is termed silvology. Silviculture also focuses on making sure that the treatments of forest stands are used to conserve and improve their productivity.

Generally, silviculture is the science and art of growing and cultivating forest crops, based on knowledge of silvics (The study of the life-history and general characteristics of Forest trees and stands, with particular reference to local/regional factors). In specific, silviculture is the practice of controlling the establishment and management of forest stands.

The distinction between forestry and silviculture is that silviculture is applied at the stand-level, while forestry is a broader concept. Adaptive management is common in silviculture, while forestry can include natural/conserved land without stand-level management and treatments being applied.

Silvicultural Systems

The origin of forestry in German-speaking Europe has defined silvicultural systems broadly as high forest, coppice with standards and compound coppice, short rotation coppice, and coppice. There are other systems as well. These varied silvicultural systems include several harvesting methods, which are often wrongly said to be a silvicultural systems, but may also be called rejuvenating or regenerating method depending on the purpose.

The high forest system is further subdivided in German:

- High forest: Age class forest.
    - Even-aged forestry.
        - Clear cutting.
        - Shelterwood cutting.
        - Seed-tree method.
    - Uneven-aged forestry.
        - The Femel selection cutting (group selection cutting).
        - Strip selection cutting (strip-and-group felling system).
        - Shelterwood wedge cutting.
        - Mixed-form regeneration methods.
- Continuous cover forestry.
    - Uneven-aged forestry.
        - Selection forest.
        - Target diameter harvesting.

These names give the impression is that these are neatly defined systems, but in practice there are variations within these harvesting methods in accordance with to local ecology and site conditions. While location of an archetypal form of harvesting technique can be identified (they all originated somewhere with a particular forester, and have been described in the scientific literature), and broad generalizations can be made, these are merely rules of thumb rather than strict blueprints on how techniques might be applied. This misunderstanding has meant that many older English textbooks did not capture the true complexity of silviculture as practiced where it originated in Mitteleuropa.

This silviculture was culturally predicated on wood production in temperate and boreal climates and did not deal with tropical forestry. The misapplication of this philosophy to those tropical forests has been problematic. There is also an alternative silvicultural tradition which developed in Japan and thus created a different biocultural landscape called satoyama.

After harvesting comes regeneration, which may be split into natural and artificial, and tending, which includes release treatments, pruning, thinning and intermediate treatments. It is conceivable that any of these 3 phases (harvesting, regeneration, and tending) may happen at the same time within a stand, depending on the goal for that particular stand.

Regeneration

Regeneration is basic to the continuation of forested, as well as to the afforestation of treeless land. Regeneration can take place through self-sown seed ("natural regeneration"), by artificially sown seed, or by planted seedlings. In whichever case, the performance of regeneration depends on its growth potential and the degree to which its environment allows the potential to be expressed. Seed, of course, is needed for all regeneration modes, both for natural or artificial sowing and for raising planting stock in a nursery.

Natural regeneration is a "human-assisted natural regeneration" means of establishing a forest age class from natural seeding or sprouting in an area after harvesting in that area through selection cutting, shelter (or seed-tree) harvest, soil preparation, or restricting the size of a clear-cut stand to secure natural regeneration from the surrounding trees.

The process of natural regeneration involves the renewal of forests by means of self-sown seeds, root suckers, or coppicing. In natural forests, conifers rely almost entirely on regeneration through seed. Most of the broadleaves, however, are able to regenerate by the means of emergence of shoots from stumps (coppice) and broken stems.

Seedbed Requirements

Any seed, self-sown or artificially applied, requires a seedbed suitable for securing germination.

In order to germinate, a seed requires suitable conditions of temperature, moisture, and aeration. For seeds of many species, light is also necessary, and facilitates the germination of seeds in other species, but spruces are not exacting in their light requirements, and will germinate without light. White spruce seed germinated at 35 °F (1.7 °C) and 40 °F (4.4 °C) after continuous stratification for one year or longer and developed radicles less than 6 cm (2.4 in) long in the cold room. When exposed to light, those germinants developed chlorophyll and were normally phototropic with continued elongation.

For survival in the short and medium terms, a germinant needs: a continuing supply of moisture; freedom from lethal temperature; enough light to generate sufficient photosynthate to support respiration and growth, but not enough to generate lethal stress in the seedling; freedom from browsers, tramplers, and pathogens; and a stable root system. Shade is very important to the survival of young seedlings. In the longer term, there must be an adequate supply of essential nutrients and an absence of smothering.

In undisturbed forest, decayed windfallen stemwood provides the most favorable seedbed for germination and survival, moisture supply being dependable, and the elevation of seedlings somewhat above the general level of the forest floor reduces the danger of smothering by leaves and snow-pressed minor vegetation; nor is such a microsite likely to be subject to flooding. Advantages conferred by those microsites include: more light, higher temperatures in the rooting zone, and better mycorrhizal development. In a survey in the Porcupine Hills, Manitoba, 90% of all spruce seedlings were rooted in rotten wood.

Mineral soil seedbeds are more receptive than the undisturbed forest floor, and are generally moister and more readily rewetted than the organic forest floor. However, exposed mineral soil, much more so than organic-surfaced soil, is subject to frost heaving and shrinkage during drought. The forces generated in soil by frost or drought are quite enough to break roots.

The range of microsites occurring on the forest floor can be broadened, and their frequency and distribution influenced by site preparation. Each microsite has its own microclimate. Microclimates near the ground are better characterized by vapour pressure deficit and net incident radiation, rather than the standard measurements of air temperature, precipitation, and wind pattern.

Aspect is an important component of microclimate, especially in relation to temperature and moisture regimes. Germination and seedling establishment of Engelmann spruce were much better on north than on south aspect seedbeds in the Fraser Experimental Forest, Colorado; the ratios of seeds to 5-year-old seedlings were determined as 32:1, 76:1, and 72:1 on north aspect bladed-shaded, bladed-unshaded, and undisturbed-shaded seedbeds, respectively. Clearcut openings of 1.2 to 2.0 hectares (3.0 to 4.9 acres) adjacent to an adequate seed source, and not more than 6 tree-heights wide, could be expected to secure acceptable regeneration (4,900, 5-year-old trees per hectare), whereas on undisturbed-unshaded north aspects, and on all seedbed treatments tested on south aspects, seed to seedling ratios were so high that the restocking of any clearcut opening would be questionable.

At least seven variable factors may influence seed germination: seed characteristics, light, oxygen, soil reaction (pH), temperature, moisture, and seed enemies. Moisture and temperature are the most influential, and both are affected by exposure. The difficulty of securing natural regeneration of Norway spruce and Scots pine in northern Europe led to the adoption of various forms of reproduction cuttings that provided partial shade or protection to seedlings from hot sun and wind. The main objective of echeloned strips or border-cuttings with northeast exposure was to protect regeneration from overheating, and was originated in Germany and deployed successfully by A. Alarik in 1925 and others in Sweden. On south and west exposures, direct insolation and heat reflected from tree trunks often result in temperatures lethal to young seedlings, as well as desiccation of the surface soil, which inhibits germination. The sun is less injurious on eastern exposures because of the lower temperature in the early morning, related to higher humidity and presence of dew.

In 1993, Henry Baldwin, after noting that summer temperatures in North America are often higher than those in places where border-cuttings have been found useful, reported the results of a survey of regeneration in a stand of red spruce plus scattered white spruce that had been isolated by clearcutting on all sides, so furnishing an opportunity for observing regeneration on different exposures in this old-field stand at Dummer, New Hampshire. The regeneration included a surprisingly large number of balsam fir seedlings from the 5% stand component of that species. The maximum density of spruce regeneration, determined 4 rods (20 m) inside from the edge of the stand on a north 20 °E exposure, was 600,000/ha, with almost 100,000 balsam fir seedlings.

A prepared seedbed remains receptive for a relatively short period, seldom as long as 5 years, sometimes as short as 3 years. Seedbed receptivity on moist, fertile sites decreases with particular rapidity, and especially on such sites, seedbed preparation should be scheduled to take advantage of good seed years. In poor seed years, site preparation can be carried out on mesic and drier sites with more chance of success, because of the generally longer receptivity of seedbeds there than those on moister sites. Although an indifferent seed year can suffice if seed distribution is good and environmental conditions favourable to seedling germination and survival, small amounts of seed are particularly vulnerable to depredation by small mammals. Considerable flexibility is possible in timing site preparation to coincide with cone crops. Treatment can be applied either before any logging takes place, between partial cuts, or after logging. In cut and leave strips, seedbed preparation can be carried out as a single operation, pre-scarifying the leave strips, post-scarifying the cut strips.

Broadcast burning is not recommended as a method of preparing sites for natural regeneration, as it rarely exposes enough mineral soil to be sufficiently receptive, and the charred organic surfaces are a poor seedbed for spruce. A charred surface may get too hot for good germination and may delay germination until fall, with subsequent over-winter mortality of unhardened seedlings. Piling and burning of logging slash, however, can leave suitable exposures of mineral soil.

Artificial Regeneration

With a view to reducing the time needed to produce planting stock, experiments were carried out with white spruce and three other coniferous species from Wisconsin seed in the longer, frost-free growing season in Florida, 125 vs. 265 days in central Wisconsin and northern Florida, respectively. As the species studied are adapted to long photoperiods, extended daylengths of 20 hours were applied in Florida. Other seedlings were grown under extended daylength in Wisconsin and with natural daylength in both areas. After two growing seasons, white spruce under long days in Florida were about the same as those in Wisconsin, but twice as tall as plants under natural Wisconsin photoperiods. Under natural days in Florida, with the short local photoperiod, white spruce was severely dwarfed and had a low rate of survival. Black spruce responded similarly. After two growing seasons, long day plants of all 4 species in Florida were well balanced, with good development of both roots and shoots, equaling or exceeding the minimum standards for 2+1 and 2+2 outplanting stock of Lake States species. Their survival when lifted in February and outplanted in Wisconsin equalled that of 2+2 Wisconsin-grown transplants. Artificial extension of the photoperiod in the northern Lake States greatly increased height increment of white and black spruces in the second growing season.

Optimum conditions for seedling growth have been determined for the production of containerized planting stock. Alternating day/night temperatures have been found more suitable than a constant temperature; at 400 lumens/m^2 light regime, a 28 °C/20 °C day/night temperatures have been recommended for white spruce. However, temperature optima are not necessarily the same at different ages and sizes. In 1984, R. Tinus investigated the effects of combinations of day and night temperature on height, caliper, and dry weight of 4 seed sources of Engelmann spruce. The 4 seed sources appeared to have very similar temperature requirements, with night optima about the same of slightly lower than daylight optima.

Tree provenance is important in artificial regeneration. Good provenance takes into account suitable tree genetics and a good environmental fit for planted / seeded trees in a forest stand. The wrong genotype can lead to failed regeneration, or poor trees that are prone to pathogens and undesired outcomes.

Artificial regeneration has been a more common method involving planting because it is more dependable than natural regeneration. Planting can involve using seedlings (from a nursery), unrooted cuttings, or seeds.

Whichever method is chosen it can be assisted by tending techniques also known as intermediate stand treatments.

The fundamental genetic consideration in artificial regeneration is that seed and planting stock must be adapted to the planting environment. Most commonly, the method of managing seed and stock deployment is through a system of defined seed zones, within which seed and stock can be moved without risk of climatic maladaptation. Ontario adopted a seed zone system in the 1970s based on G.A. Hills' 1952 site regions and provincial resource district boundaries, but Ontario's seed zones are now based on homogeneous climatic regions developed with the Ontario Climate Model. The regulations stipulate that source-identified seedlots may be either a general collection, when only the seed zone of origin is known, or a stand collection from a specific latitude and longitude. The movement of general-collection seed and stock across seed zone boundaries is prohibited, but the use of stand-collection seed and stock in another seed zone is acceptable when the Ontario Climate Model shows that the planting site and place of seed origin are climatically similar. The 12 seed zones for white spruce in Quebec are based mainly on ecological regions, with a few modifications for administrative convenience.

Seed quality varies with source. Seed orchards produce seed of the highest quality, then, in order of decreasing seed quality produced, seed production areas and seed collection areas follow, with controlled general collections and uncontrolled general collections producing the least characterized seed.

Seeds

Dewinging and Extraction

When seed is first separated from cones it is mixed with foreign matter, often 2 to 5 times the volume of the seed. The more or less firmly attached membranous wings on the seed must be detached before it is cleaned of foreign matter. The testa must not incur damage during the dewinging process. Two methods have been used, dry and wet. Dry seed may be rubbed gently through a sieve that has a mesh through which only seed without wings can pass. Large quantities of seed can be processed in dewinging machines, which use cylinders of heavy wire mesh and rapidly revolving stiff brushes within to remove the wings. In the wet process, seed with wings attached are spread out 10 cm to 15 cm deep on a tight floor and slightly moistened throughout; light leather flails are used to free seed from the wings. B. Wang described a unique wet dewinging procedure in 1973 using a cement mixer, used at the Petawawa tree seed processing facility. Wings of white and Norway spruce seed can be removed by dampening the seed slightly before it is run through a fanning mill for the last time. Any moistened seed must be dried before fermentation or moulding sets in.

Seed Viability

A fluorescein diacetate (FDA) biochemical viability test for several species of conifer seed, including white spruce, estimates the proportion of live seed (viability) in a seedlot, and hence the percentage germination of a seedlot. The accuracy of predicting percentage germination was within +/- 5 for most seedlots. White spruce seed can be tested for viability by an indirect method, such as the fluorescein diacetate (FDA) test or 'Ultra-sound'; or by the direct growth method of 'germination'. Samples of white spruce seed inspected in 1928 varied in viability from 50% to 100%, but averaged 93%. A 1915 inspection reported 97% viability for white spruce seed.

Germinative Testing

The results of a germination test are commonly expressed as germinative capacity or a germination percentage, which is the percentage of seeds that germinate during a period of time, ending when germination is practically complete. During extraction and processing, white spruce seeds gradually lost moisture, and total germination increased. Mittal et al. reported that white spruce seed from Algonquin Park, Ontario, obtained the maximum rate (94% in 6 days) and 99% total germination in 21 days after 14-week pre-chilling. The pre-treatment of 1% sodium hypochlorite increased germinability.

Encouraged by Russian success in using ultrasonic waves to improve the germinative energy and percentage germination of seeds of agricultural crops, Timonin demonstrated benefits to white spruce germination after exposure of seeds to 1, 2, or 4 minutes of ultrasound generated by an M.S.E. ultrasonic disintegrator with a power consumption of 280 VA and power impact of 1.35 amperes. However, no seeds germinated after 6 minutes of exposure to ultrasound.

Seed Dormancy

Seed dormancy is a complex phenomenon and is not always consistent within species. Cold stratification of white spruce seed to break dormancy has been specified as a requirement, but Heit and Hellum regarded stratification as unnecessary. Cone handling and storage conditions affect dormancy in that cold, humid storage (5 °C, 75% to 95% relative humidity) of the cones prior to extraction seemingly eliminated dormancy by overcoming the need to stratify. Periods of cold, damp weather during the period of cone storage might provide natural cold (stratification) treatment. Once dormancy was removed in cone storage, subsequent kiln-drying and seed storage did not reactivate dormancy.

Haddon and Winston found a reduction in viability of stratified seeds after 2 years of storage and suggested that stress might have been caused by stratification, e.g., by changes in seed biochemistry, reduced embryo vigor, seed aging or actual damage to the embryo. They further questioned the quality of the 2-year-old seed even though high germination occurred in the samples that were not stratified.

Cold Stratification

Cold stratification is the term applied to the storing of seeds in (and, strictly, in layers with) a moist medium, often peat or sand, with a view to maintaining viability and overcoming dormancy. Cold stratification is the term applied to storage at near-freezing temperatures, even if no medium is used. A common method of cold stratification, is to soak seed in tap water for up to 24 h, superficially dry it, then store moist for some weeks or even months at temperatures just above freezing. Although Hellum found that cold stratification of an Alberta seed source led to irregular germination, with decreasing germination with increasing length of the stratification period, Hocking's paired test with stratified and nonstratified Alberta seed from several sources revealed no trends in response to stratification. Hocking suggested that seed maturity, handling, and storage needed to be controlled before the need for stratification could be determined. Later, Winston and Haddon found that the storage of white spruce cones for 4 weeks at 5 °C prior to extraction obviated the need for stratification.

Seed Ripeness

Seed maturity cannot be predicted accurately from cone flotation, cone moisture content, cone specific gravity; but the province of B.C. found embryo occupying 90% + of the corrosion cavity and megagametophyte being firm and whitish in colour are the best predictors for white spruce in B.C., and Quebec can forecast seed maturity some weeks in advance by monitoring seed development in relation to heat-sums and the phenological progression of the inflorescence of fireweed (Epilobium angustifolium L.), an associated plant species. Cone collection earlier than one week before seed maturity would reduce seed germination and viability during storage. Four stages of maturation were determined by monitoring carbohydrates, polyols, organic acids, respiration, and metabolic activity. White spruce seeds require a 6-week post-harvest ripening period in the cones to obtain maximum germinability, however, based on cumulative degree-days, seed from the same trees and stand showed 2-week cone storage was sufficient.

Forest Tree Plantations

Plantation Establishment Criteria

Plantations may be considered successful when outplant performance satisfies certain criteria. The term "free growing" is applied in some jurisdictions. Ontario's "Free-to-Grow" (FTG) equivalent relates to a forest stand that meets a minimum stocking standard and height requirement, and is essentially free of competition from surrounding vegetation that might impede growth. The FTG concept was introduced with the advent of the Forest Management Agreement program in Ontario in 1980 and became applicable to all management units in 1986. Policy, procedures, and methodologies readily applicable by forest unit managers to assess the effectiveness of regeneration programs were still under development during the Class Environmental Assessment hearings.

In British Columbia, the Forest Practices Code governs performance criteria. To minimize the subjectivity of assessing deciduous competition as to whether or not a plantation is established, minimum specifications of number, health, height, and competition have been specified in British Columbia. However, minimum specifications are still

subjectively set and may need to be fine-tuned in order to avoid unwarranted delay in according established status to a plantation. For example, a vigorous white spruce with a strong, multi-budded leading shoot and its crown fully exposed to light on 3 sides would not qualify as free-growing in the current British Columbia Code but would hardly warrant description as unestablished.

Competition

Competition arises when individual organisms are sufficiently close together to incur growth constraint through mutual modification of the local environment. Plants may compete for light, moisture and nutrients, but seldom for space per se. Vegetation management directs more of the site's resources into usable forest products, rather than just eliminating all competing plants. Ideally, site preparation ameliorates competition to levels that relieve the outplant of constraints severe enough to cause prolonged check.

The diversity of boreal and sub-boreal broadleaf-conifer mixed tree species stands, commonly referred to as the "mixedwoods", largely preclude the utility of generalizations and call for the development of management practices incorporating the greater inherent complexity of broadleaf-conifer mixtures, relative to single-species or mixed-species conifer forest. After harvesting or other disturbance, mixedwood stands commonly enter a prolonged period in which hardwoods overtop the coniferous component, subjecting them to intense competition in an understorey. It is well established that the regeneration and growth potential of understorey conifers in mixedwood stands is correlated to the density of competing hardwoods. To help apply "free-to-grow" regulations in British Columbia and Alberta, management guidelines based on distance-dependent relations within a limited radius of crop trees were developed, but Lieffers et al. found that free-growing stocking standards did not adequately characterize light competition between broadleaf and conifer components in boreal mixedwood stands, and further noted that adequate sampling using current approaches would be operationally prohibitive.

Many promising plantations have failed through lack of tending. Young crop trees are often ill-equipped to fight it out with competition resurgent following initial site preparation and planting.

Perhaps the most direct evaluation of the effect of competition on plantation establishment is provided by an effective herbicide treatment. The fact that herbicide treatment does not always produce positive results should not obscure the demonstrated potential of herbicides for significantly promoting plantation establishment. Factors that can vitiate the effectiveness of a herbicide treatment include: weather, especially temperature, prior to and during application; weather, especially wind, during application; weather, especially precipitation, in the 12 to 24 hours after application; vegetation characteristics, including species, size, shape, phenological stage, vigour, and distribution of weeds; crop characteristics, including species, phenology, and condition, the effects of other treatments, such as preliminary shearblading, burning or other prescribed or accidental site preparation; and the herbicide used, including dosage, formulation, carrier, spreader, and mode of application. There is a lot that can go wrong, but a herbicide treatment can be as good or better than any other method of site preparation.

Competition Indices

The study of competition dynamics requires both a measure of the competition level and a measure of crop response. Various competition indices have been developed, e.g., by Bella and Hegyi based on stem diameter, by Arney, Ek and Monserud, and Howard and Newton based on canopy development, and Daniels, Wagner, and Weiner with proximity-based models. Studies generally considered tree response to competition in terms of absolute height or basal area, but Zedaker and Brand sought to quantify crop tree size and environmental influences by using relative growth measures.

Tending

Tending is the term applied to pre-harvest silvicultural treatment of forest crop trees at any stage after initial planting or seeding. The treatment can be of the crop itself (e.g., spacing, pruning, thinning, and improvement cutting) or of competing vegetation (e.g., weeding, cleaning).

Planting

How many trees per unit area (spacing) that should be planted is not an easily answered question. Establishment density targets or regeneration standards have commonly been based on traditional practice, with the implicit aim of getting the stand quickly to the free-to-grow stage. Money is wasted if more trees are planted than are needed to achieve desired stocking rates, and the chance to establish other plantations is proportionately diminished. Ingress (natural regeneration) on a site is difficult to predict and often becomes surprisingly evident only some years after planting has been carried out. Early stand development after harvesting or other disturbance undoubtedly varies greatly among sites, each of which has its own peculiar characteristics.

For all practical purposes, the total volume produced by a stand on a given site is constant and optimum for a wide range of density or stocking. It can be decreased, but not increased, by altering the amount of growing stock to levels outside this range. Initial density affects stand development in that close spacing leads to full site utilization more quickly than wider spacing. Economic operability can be advanced by wide spacing even if total production is less than in closely spaced stands.

Beyond the establishment stage, the relationship of average tree size and stand density is very important. Various density-management diagrams conceptualizing the density-driven stand dynamics have been developed. Smith and Brand's diagram has mean tree volume on the vertical axis and the number of trees/ha on the horizontal axis: a stand can either have many little trees or a few big ones. The self-thinning line shows the largest number of trees of a given size/ha that can be carried at any given time. However, Willcocks and Bell caution against using such diagrams unless specific knowledge of the stand trajectory is known.

In the Lake States, plantations have been made with the spacing between trees varying from 3 by 3 to 10 by 10 feet (0.9 m by 0.9 m to 3.0 m by 3.0 m). Kittredge recommended that no fewer than 600 established trees per acre (1483/ha) be present during the early life of a plantation. To insure this, at least 800 trees per acre (1077/ha) should be planted where 85% survival may be expected, and at least 1200/ac (2970/ha) if only half of them can be expected to live. This translates into recommended spacings of 5 by 5 to 8 by 8 feet (1.5 m by 1.5 m to 2.4 m by 2.4 m) for plantings of conifers, including white spruce in the Lake States.

Enrichment Planting

A strategy for enhancing natural forests' economic value is to increase their concentration of economically important, indigenous tree species by planting seeds or seedlings for future harvest, which can be accomplished with enrichment planting (EP). This means increasing the planting density (i.e., the numbers of plants per hectare) in an already growing forest stand."

Release Treatments

- Weeding: A process of getting rid of saplings' or seedlings' competition by mowing, application of herbicide, or other method of removal from the surroundings.
- Cleaning: Release of select saplings from competition by overtopping trees of a comparable age. The treatment favors trees of a desired species and stem quality.
- Liberation cutting: A treatment that releases tree seedling or saplings by removing older overtopping trees.

Spacing

Over-crowded regeneration tends to stagnate. The problem is aggravated in species that have little self-pruning ability, such as white spruce. Spacing is a thinning (of natural regeneration), in which all trees other than those selected for retention at fixed intervals are cut. The term juvenile spacing is used when most or all of the cut trees are unmerchantable. Spacing can be used to obtain any of a wide range of forest management objectives, but it is especially undertaken to reduce density and control stocking in young stands and prevent stagnation, and to shorten the rotation, i.e., to speed the production of trees of a given size. Volume growth of individual trees and the merchantable growth of stands are increased. The primary rationale for spacing is that thinning is the projected decline in maximum allowable cut. And since wood will be concentrated on fewer, larger, and more uniform stems, operating and milling costs will be minimized.

Methods for spacing may be: manual, using various tools, including power saws, brush saws, and clippers; mechanical, using choppersand mulchers; chemical; or combinations of several methods. One treatment has had notable success in spacing massively overstocked (<100 000 stems/ha) natural regeneration of spruce and fir in Maine. Fitted to helicopter, the Thru-Valve boom emits herbicide spray droplets 1000 μm to 2000 μm in diameter at very low pressure. Swaths 1.2 m wide and leave strips 2.4 m wide were obtained with "knife-edge" precision when the herbicide was applied by helicopter flying at a height of 21 m at a speed of 40–48 km/h. It seems likely that no other method could be as cost-effective.

Twenty years after spacing to 2.5 × 2.5 m, 30-year-old mixed stands of balsam fir and white spruce in the Green River watershed, New Brunswick, averaged 156.9 m^3/ha.

A spacing study of 3 conifers (white spruce, red pine and jack pine) was established at Moodie, Manitoba, on flat, sandy, nutritionally poor soils with a fresh moisture regime. Twenty years after planting, red pine had the largest average dbh, 15% greater than jack pine, while white spruce dbh was less than half that of the pines. Crown width showed a gradual increase with spacing for all 3 conifers. Results to date were suggesting optimum spacings between 1.8 m and 2.4 m for both pines; white spruce was not recommended for planting on such sites.

Comparable data are generated by espacement trials, in which trees are planted at a range of densities. Spacings of 1.25 m, 1.50 m, 1.75 m, 2.00 m, 2.50 m, and 3.00 m on 4 site classes were used in the 1922 trial at Petawawa, Ontario. In the first of 34 old field white spruce plantations used to investigate stand development in relation to spacing at Petawawa, Ontario, regular rows were planted at average spacings of from 4 × 4 to 7 × 7 feet (1.22 m × 1.22 m to 2.13 m × 2.13 m). Spacings up to 10 × 10 feet (3.05 m × 3.03 m) were subsequently included in the study. Yield tables based on 50 years of data showed:

- Except for merchantable volumes at age 20 and site classes 50 and 60, closer spacings gave greater standing volumes at all ages than did wider spacings, the relative difference decreasing with age.
- Merchantable volume as a proportion of total volume increases with age, and is greater at wider than at closer spacings.
- Current annual volume increment culminates sooner at closer than at wider spacings.

A smaller espacement trial, begun in 1951 near Thunder Bay, Ontario, included white spruce at spacings of 1.8 m, 2.7 m, and 3.6 m. At the closest spacing, mortality had begun at 37 years, but not at the wider spacings.

The oldest interior spruce espacement trial in British Columbia was established in 1959 near Houston in the Prince Rupert Forest Region. Spacings of 1.2 m, 2.7 m, 3.7 m, and 4.9 m were used, and trees were measured 6, 12, 16, 26, and 30 years after planting. At wide espacements, trees developed larger diameters, crowns, and branches, but (at 30 years) basal area and total volume/ha were greatest in the closest espacement. In more recent trials in the Prince George Region of British Columbia and in Manitoba, planting density of white spruce had no effect on growth after up to 16 growing seasons, even at spacings as low as 1.2 m. The slowness of juvenile growth and of crown closure delay the response to intra-competition. Initially, close spacing might even provide a positive nurse effect to offset any negative response to competition.

Thinning

Thinning is an operation that artificially reduces the number of trees growing in a stand with the aim of hastening the development of the remainder. The goal of thinning is to control the amount and distribution of available growing space. By altering stand density, foresters can influence the growth, quality, and health of residual trees. It also provides an opportunity to capture mortality and cull the commercially less desirable, usually smaller and malformed, trees. Unlike regeneration treatments, thinnings are not intended to establish a new tree crop or create permanent canopy openings.

Thinning greatly influences the ecology and micro-meteorology of the stand, lowering the inter-tree competition for water. The removal of any tree from a stand has repercussions on the remaining trees both above-ground and below. Silvicultural thinning is a powerful tool that can be used to influence stand development, stand stability, and the characteristics of the harvestable products.

When considering intensive conifer plantations designed for maximum production, it is essential to remember that tending and thinning regimes and wind and snow damage are intimately related.

Previous studies have demonstrated that repeated thinnings over the course of a forest rotation increase carbon stores relative to stands that are clear-cut on short rotations and that the carbon benefits differ according to thinning method.

Precommercial Thinning

In the early development of forest stand, density of trees remain high and there is competition among trees for nutrients. When natural regeneration or artificial seeding has resulted in dense, overstocked young stands, natural thinning will in most cases eventually reduce stocking to more silviculturally desirable levels. But by the time some trees reach merchantable size, others will be overmature and defective, and others will still be unmerchantable. To reduce this unbalance and to obtain more economic returns, in the early stage, one kind of cleaning is done which is known as precommercial thinning. Generally, one or two times precommercial thinning is done to facilitate the growth of the tree The yield of merchantable wood can be greatly increased and the rotation shortened by precommercial thinning. Mechanical and chemical methods have been applied, but their costliness has militated against their ready adoption.

Pruning

Pruning, as a silvicultural practice, refers to the removal of the lower branches of the young trees (also giving the shape to the tree) so clear knot-free wood can subsequently grow over the branch stubs. Clear knot-free lumber has a higher value. Pruning has been extensively carried out in the Radiata pine plantations of New Zealand and Chile, however the development of Finger joint technology in the production of lumber and mouldings has led to many forestry companies reconsidering their pruning practices. "Brashing" is an alternative name for the same process. Pruning can be done to all trees, or more cost effectively to a limited number of trees. There are two types of pruning: natural or self-pruning and artificial pruning. Most cases of self-pruning happen when branches do not receive enough sunlight and die. Wind can also take part in natural pruning which can break branches. Artificial pruning is where people are paid to come and cut the branches. Or it can be natural, where trees are planted close enough that the effect is to cause self-pruning of low branches as energy is put into growing up for light reasons and not branchiness.

Stand Conversion

The term stand conversion refers to a change from one silvicultural system to another and includes species conversion, i.e., a change from one species (or set of species) to another. Such change can be effected intentionally by various silvicultural means, or incidentally by default e.g., when high-grading has removed the coniferous content from a mixedwood stand, which then becomes exclusively self-perpetuating aspen. In general, such sites as these are the most likely to be considered for conversion.

NATURAL REGENERATION

Regeneration (reproduction) refers to the seedlings or saplings existing in a stand. In silviculture, it is the act of renewing tree cover by establishing young trees naturally or artificially. Natural regeneration is the establishment of young trees through natural seeding, sprouting, suckering, or layering. Artificial regeneration is the establishment of young trees through planting or direct seeding.

A regeneration method is a procedure by which a stand is established or renewed by means of natural or artificial reproduction. The various methods include the removal of the old stand or cohort (usually involving a harvest), the establishment of a new one, and any supplementary treatments of vegetation, slash, or soil that are applied to create conditions favorable to the establishment of reproduction. A regeneration cutting is any removal of trees intended to assist regeneration already present or to make regeneration possible.

Stand age structure is an important characteristic used to characterize and classify stands and silvicultural systems. The three basic classified age structures are:

- Even-aged: A stand where the trees have only small differences in their ages (a single age class). By convention, the spread of ages does not differ by more than 20% of the intended rotation.
- Two-aged: A stand with trees of two distinct age classes, separated in age by more than 20% of rotation.
- Uneven-aged: A stand where the trees differ markedly in their ages, with trees of three or more distinct age classes either mixed or in small groups.

Silvicultural systems and regeneration methods can be classified in order to systematize the description of the wide variety of procedures used in practice. The following stand level (generally > 2 acres) classification of planned natural regeneration methods is based on the mode of origin of regeneration and on the arrangement of cuttings in space and time. Gradations and hybrids between methods are inherent to the classification. Reserve trees can be included with any of the following even-aged methods.

Even-aged Systems

- Coppice Method:
  - Simple Coppice.
  - Coppice with Standards.
  - Compound Coppice.
- Clearcut Method:
  - Uniform Clearcut.
  - Alternate (strip, patch) Clearcut.
  - Progressive (strip, patch) Clearcut.

- Seed Tree Method:
    - Single Seed Tree.
    - Group Seed Tree.
- Overstory Removal Method:
    - Uniform Overstory Removal.
    - Patch Overstory Removal.
- Shelterwood Method:
    - Uniform Shelterwood.
    - Strip Shelterwood.
    - Patch Shelterwood.

Uneven-aged Systems

- Single-Tree (Gap) Selection Method.
- Group Selection Method.
- Patch Selection Method.

The cover type chapters can be referenced for specific recommended regeneration methods and silvicultural systems. In general, the discussion of regeneration methods assumes that the stand management objectives include maintaining the current cover type, regenerating the stand promptly following harvest, and promoting stand vigor and health. When interpreting natural regeneration methods to encourage cover type conversion, refer to the guidelines for the target cover type; an adequate seed source or sprout/sucker stock must be available. When managing mixed stands, refer to guidelines for each individual cover type and adapt techniques that will facilitate each, either in patches or intermingled. Following the implementation of a regeneration method, it is necessary to assess the adequacy of regeneration and to determine the need for additional treatments.

In application, silvicultural systems are developed based predominantly on consideration of the silvical characteristics of the forest cover type, site potentials, and landowner goals. Stands and sites tend to be heterogeneous units and require adaptive interpretation and management. As stand and site characteristics vary, so do current management alternatives and potentials to meet different management objectives. General methods and systems can provide a guide to the development of stand level prescriptions, but must be sufficiently flexible to respond to variable stand and site conditions and to facilitate adaptive silviculture to meet landowner goals and objectives.

COPPICE METHOD

A silvicultural method designed to naturally regenerate a stand using vegetative reproduction. Stump or root sprouts form the majority of the new stand although some seed origin reproduction may be present.

Characteristics

- The basic method results in even-aged stands; however, the variations can result in two-aged or multi-aged stands.
- Vegetative reproduction (low forest).
- Reproduction from stump sprouts or root suckers.
- Potentially applicable to most hardwood species.
- New stand regenerates after the existing stand is harvested.

Contrast with other Methods

Differs from other even-aged regeneration methods (seed tree, shelterwood, overstory removal, and clearcut) in that the regenerated stand is derived from vegetative reproduction rather than a seed source. Generally, there is no residual stand left as the residual can interfere with regeneration and is not necessary to shelter the regenerated stand. It differs from selection in that it is an even-aged system and the reproduction is not from seed. The method somewhat mimics catastrophic loss and replacement of a stand, but differs in the amount and distribution of residual trees and coarse woody debris.

Variations to this Method

- Simple coppice: A complete harvest is applied to the entire stand to be regenerated. This variation produces a stand similar in composition to the parent stand except that any conifer component will be reduced. Results in even-aged stand.
- Coppice with Standards: A complete harvest is applied to the entire stand with the exception of chosen standards. The standards might be crop trees of the same species as the stand to be harvested but are often a different species. This variation can be used to slowly convert a stand to a different composition, to develop a seed-origin stand, or to manage mixed stands. It can result in a two-aged stand depending on the number of residual trees.
- Compound Coppice: A modification of the coppice with standards method that incorporates two or more age classes of standards above the coppice stand. The age classes of the standards will be multiples of the rotation age of the coppiced stand. Results in two-aged or uneven-aged stand.

Application

General considerations in the application of the coppice method are:

- Spacing: For species reproducing from stump sprouts, spacing of the regenerated stand is determined by the spacing of the parent stand. For root suckering species, spacing of the parent stand is not as important of a factor.
- Sprouting capability: The vigor of the sprouting response varies with age and size of the parent trees. This response is species-specific and highly variable. In general, sprouting is most vigorous in trees that are experiencing their most rapid growth.
- Cutting season - Site capability.
- Competition.
- Overstory composition, condition, and health.

Advantages and Disadvantages

Coppice with standards regeneration method.

Advantages:

- Simple, dependable regeneration.
- Efficiency of harvesting operations.
- No site preparation is needed - No delay in regeneration.
- Growing space continuously occupied - Longer time period between entries reduces some vehicle impacts to soils.

Disadvantages:

- Spacing of stump sprouts may contribute to poorly.

- formed trees and an understocked stand.
- Potential for water table changes on wet sites.
- Higher windthrow potential (standards, adjacent stand).

CLEARCUT METHOD

A method of regenerating an even aged stand in which a new age class develops in a fully exposed microclimate after removal, in a single cutting, of all trees in the previous stand. Regeneration is from natural seeding, direct seeding, planted seedlings, and/or advance reproduction. The management unit or stand in which regeneration, growth, and yield are regulated consists of the individual clearcut stand.

A 120 acre clearcut on a xeric ridgetop in East Texas prior to site preparation and re-planting. The previous and next rotations are both loblolly pine plantations. Notice that there is little slash or debris apparent on the site, a common feature of clearcut plantations of coniferous species.

A large commercial clearcut of a bottomland hardwood stand in the Coastal Plain. Notice that there is much more slash apparent, as hardwoods tend to have wider crowns and larger branches than softwoods. The retained large trees are nonmerchantable due to poor form, rot, or their species. They will either be left through the next rotation, or felled as part of site preparation activities.

Extent

Clearcutting is the most common regeneration method used in North America today, particularly in the US South. It is well-suited to plantation forests and many naturally regenerated stands.

Cuts Involved

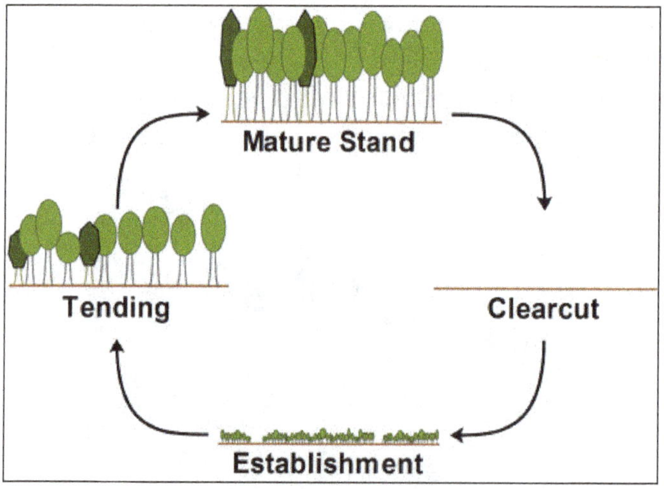

An example of the clearcut regeneration method. Note that this is a one-harvest system that regenerates an even-aged stand.

Clearcutting is so commonly applied in part due to its simplicity. Clearcutting involves only a single cut that removes all merchantable timber to regenerate the stand. Following this cut, the stand is either artificially (e.g. planting or direct seeding) or naturally regenerated with trees of a single age class.

A three-year-old clearcut of a bottomland hardwood stand in the US South shows how rapidly a new cohort naturally establishes itself, even without tree planting in this case.

An approximately 20-year-old clearcut regenerated with western white pine planted at various densities. In some cases a stand will not adequately regenerate following clear-cutting without appropriate site preparation and artificial regeneration. In such cases it is the silviculturist's ethical obligation to ensure such activities succeed if clearcutting is prescribed.

Common Variations

A commercial clearcut is the most typical variation of the clearcut regeneration method. This involves the removal of all merchantable timber. A clearcut may thus still have standing live trees remaining following harvest. These are usually not intentionally retained (as in a two age regen method), but rather are simply not cut. There are a variety of reasons trees may not be cut.

- They may be too small to be merchantable.
- Poor stem form may limit merchantability, particularly in locations with weak pulpwood markets.
- Heart rot may make otherwise suitable trees unmerchantable.
- Some species are not useful for producing forest products.
- There may be no market at the time of harvest for trees that would usually be merchantable.

Trees left after a commercial clearcut may either be left, or treated during site preparation. During site prep trees may be deadened with herbicides or by girdling and left standing. They may also be felled with a chainsaw or shear, and then treated with the rest of the slash. When deciding what to do with unmerchantable stems, both their effect on the new cohort and the cost of handling them should be considered.

A biological clearcut is a harvest where all trees are removed, regardless of size class or merchantability. Biological clearcuts are used for research purposes, but are rarely, if ever, prescribed in operational forests.

A clearcut of a bottomland hardwood stand in the US South. The standing trees are pecans. While pecan has high value for furniture, flooring, and other products, there were no mills accepting it at the time of harvest, leading these trees to be left as unmerchantable.

A cleancut is a clearcut in which both merchantable and smaller unmerchantable trees are cut down to some minimum size limit. Cleancutting is prescribed for species where shade created by a midstory will prevent adequate regeneration. For example, southern bottomland hardwoods being managed for oaks are often cleancut, with all stems more than 2 inches dbh removed to prevent shading the new cohort.

A cleancut of a bottomland hardwood stand in the US South. Note that there are no remaining smaller trees that will shade out the new cohort.

The spatial arrangement of clearcutting can also be varied. Strip clearcuts are used when the species being regenerated are not windfirm, or when natural seeding is required for regeneration but seed dispersal distances are limited. Strip clearcuts are often done in a series. Once a clearcut strip has regenerated, the adjacent strip is then cut. Considering prevailing wind directions and orienting strips appropriately is important for either reason that strips are being utilized.

Another commonly utilized method of clearcutting is the patch clearcut. Patch clearcuts are the same as clearcutting, but are conducted on a smaller spatial scale. Patch clearcuts may be no more than a few acres in size.

A strip clearcut being applied in a spruce-fir stand in Maine to limit losses to windthrow. Spruce and fir are shallow-rooted species that often establish on low-strength, coarsely-textured soils high in organic matter, and are thus extremely susceptible to windthrow.

Ecological Considerations (Silvics)

Clearcutting results in the most fully exposed microclimate of any regeneration method. It is best suited to shade intolerant species. Many pioneer species are appropriate for regenerating via clearcutting.

A clearcut in a mixed conifer stand in California shows the shaded southern edge (left side of photo) of the clearcut contrasted by the well-lit interior and northern edge.

While clearcuts result in more light reaching ground-level, the amount and quality of light varies across a clearcut. Along with light, air temperature, soil temperature, relative humidity, and soil moisture all also vary. In the northern hemisphere, the sun is generally found in the southern portion of the sky throughout the day. Thus, the south edge of a clearcut in an area with level topography will be shaded by the adjacent stand, while the center and north edge of a clearcut will receive more light. Eastern edges are shaded in the morning when it is relatively cool, and are fully exposed to the sun in the afternoon, when it is warmer. The opposite trend applies to western edges. Edge environments are often more variable in their microclimate than the interior of a clearcut, or even an uncut stand. So although a clearcut may appear relatively homogenous, this spatial variability in the light environment may create patches suitable to the regeneration of species with varying shade tolerances.

The size, geometry, and orientation of clearcut patches will also influence the light environment. Larger clearcuts tend to receive more direct light rather than diffuse light, which can result in hotter daytime temperatures and cooler nighttime temperatures.

Beyond light, newly establishing seedlings are also affected by belowground processes, including competition for soil resources with the roots of adjacent stands and non-crop vegetation. In general, increasing root biomass of herbaceous and weedy species results in reduced growth of desirable species. Despite this effect, light often remains the most limiting resource for young seedlings compared to nutrient or water availability. Because seedlings are small, their nutrient and water demands are low compared to larger trees. This is coupled with greater nutrient and water availability following clearcutting as slash decomposes, releasing nutrients and the evapotranspirative demand of the harvested stand is greatly reduced.

As with any regeneration method, it is important to consider how the new cohort will establish on the site following clearcutting. If planting or direct seeding of a shade intolerant pioneer species is planned, then clearcut size and shape are not constrained by the need to ensure adequate regeneration. If natural regeneration from seed is desired, then the natural patterns of seed dispersal must be taken into account. For species with wind-dispersed seed, it is necessary to know the distance that seed will be distributed from surrounding stands and the prevailing wind direction. If clearcuts are too large, seed of desirable species may not reach all the harvested area, resulting in a regeneration failure.

In some species it may be possible to rely on stump or root sprouting as a source of regeneration following clearcutting. Sprouting potential varies with species, tree size, and age. If sprouting is desired, then any available data on sprouting for the species in question should be used to determine if regeneration will be sufficient following harvest.

Aspen and poplar root sprouts following a clearcut in Utah. Given the tendency of many

Populus species to prolifically root sprout, it is often possible to rely on sprouting for regeneration rather than establishment from seed or artificial means.

Protection of the soil resource is critical if silviculture is to be a sustainable practice. Many forests in the US South were established on previously eroded lands that were degraded through unsustainable agricultural practices. While topsoil may be very thin in these stands, it is not the result of past silvicultural practices. In more mountainous regions, such as the western US, the potential for erosion is higher. Mass wasting, or the bulk movement of soil as in a landslide, generally increases in both soil volume lost and frequency of slides as slopes become increasingly steep. In general, the vast majority of erosion associated with clearcutting occurs on forest roads and skid trails, not on the harvested matrix. Erosion from the harvested matrix may be higher for several years following clearcutting before vegetation can become established enough to retain the soil. Streamside management zones, or buffers of unharvested trees maintained around streams, are an effective means of preventing eroded sediment from entering streams.

Clearcuts of mixed coniferous stands in Oregon. Proper road construction and best management practice compliance is critical to prevent erosion from clearcut harvest operations on steep terrain such as this.

Clearcutting has been found to have mixed effects on various species of wildlife and their habitat. For example, clearcutting in boreal forests was found to vary in its effect on wildlife depending on the size of their home ranges. In this study, animals with large home ranges (> 2 sq. miles) continued to utilize suitable areas of clearcuts and surrounding vegetated areas. Animals with medium-sized home ranges (< 60 acres) no longer used the clearcut patches. Those with the smallest home ranges (< 2.5 acres) either remained in the clearcut area, or relocated to surrounding intact stands. Effects of clearcutting on wildlife also vary by species. For instance, clearcutting resulted in reduced abundance of some salamander species, but had no effect on others in the Pacific Northwest. Some species, such as white-tailed deer, prefer recently clearcut areas as a source of abundant browse. If management for wildlife habitat is a silvicultural objective, an understanding of the effects of clearcutting on the specific species under management is necessary.

ECOLOGICAL THINNING

Fresh snow partially covers epixylic Rough-stalked Feather-moss (Brachythecium rutabulum), growing on a dead hybrid black poplar (Populus x canadensis), thinned in 2008 to create a waterside path. The last stage of the moss lifecycle is shown, where the sporophytes are visible before dispersion of their spores:

- The calyptra is still attached to the sporangium.
- The tops of the gametophytes.
- Can be discerned as well.

Inset shows the surrounding, black poplars growing on sandy loam on the bank of a kolk, with the detail area marked.

Ecological thinning is a silvicultural technique used in forest management that involves cutting trees to improve functions of a forest other than timber production.

Although thinning originated as a man-made forest management tool, aimed at increasing timber yields, the shift from production forests to multifunctional forests brought with it the cutting of trees to manipulate an ecosystem for various reasons, ranging from removing non-native species from a plot to removing poplars growing on a riverside beach aimed at recreational use.

Since the 1970s, leaving the thinned trees on the forest floor has become an increasingly common policy: wood can be decomposed in a more natural fashion, playing an important role in increasing biodiversity by providing habitat to various invertebrates, birds and small mammals. Many fungi (e.g. Calocera viscosa) and mosses are saproxylic or epixylic as well (e.g. Marchantiophyta) – some moss species completing their entire life-cycle on a single log.

Where trees are managed under a commercial regime, competition is reduced by removing adjacent stems that exhibit less favourable timber quality potential. When left in a natural state trees will "self-thin", but this process can be unreliable in some circumstances. Examples of this can be found in the Buxus – Ironbark forests and woodlands of Victoria (Australia) where a large proportion of trees are coppice, resultant from timber cutting in decades gone by.

Ecophysiological Repercussions

Thinning decreases canopy closure and increases the penetration of solar radiation into the canopy. The photosynthetic efficiency of this energy is improved, and needle retention is prolonged, especially in the lower parts of the crown. The root system, crown length, crown diameter, and crown area all increase after thinning. Even if soil evaporation and individual tree transpiration increases after thinning, total evapo-transpiration at stand level tends to decrease; canopy water interception is reduced and throughfall increased, so that tree-water status usually improves after thinning.

Radial annual growth is an integrative index of tree physiological response to environmental variation. Working with Norway spruce, which can be expected to behave in some respects similarly to white spruce, Misson et al. showed that a reduction in stand density alters the classical climate–growth relationship. At individual tree level, thinning could be used to increase tree resistance to drought stress. Nevertheless, this effect is limited when site conditions are limiting. Misson et al. concluded that heavy thinning should be applied when forest decline is expected from drought stress. Furthermore, stands on dry sites should be thinned more heavily because such sites cannot support high-density stands.

Aussenac, also working with Norway spruce, investigated the response to thinning. Thinning altered the temporal evolution of radial growth at all frequencies in radial growth chronologies. Previous studies had shown, not surprisingly, that thinning decreases the canopy closure and encourages solar radiation to penetrate the canopy. Solar energy then becomes more important inside the crown. The improved photosynthetic efficiency of this energy influences the retention time of needles, especially in the lower part of the crown. Furthermore, the length, diameter, and area, of the crown, and the size of the root system all increase after thinning.

These modifications at the crown level influence photosynthate production positively as long as water supply is not limiting. Even if soil evaporation and individual tree transpiration are more important after thinning, total evapo-transpiration at stand level tends to decrease. Furthermore, with the reduced importance of leaf area index (LAI), thinning reduces canopy water interception and increases throughfall. This explains why the soil moisture content increases in a thinned stand. Thus, during the vegetation period, tree-water status is usually better in a thinned stand than in a high-density stand.

For semi-tolerant species such as Norway spruce and white spruce, solar energy is less limiting than for intolerant species. This can account for the exponential relationship found by Misson et al. between thinning intensity and mean radial growth. Only when thinning was relatively severe did important radial growth variation occur. Furthermore, the ecophysiological advantage of an improved water supply could be counterbalanced by limiting site conditions.

For Norway spruce in the Belgian Ardennes, Misson et al. recommended that stand basal area should not exceed 26 m^2/ha on a dry site, or 29 m^2/ha on a moist site if the objective is to maintain high long-term radial growth.

Misson et al. found that adaptations to thinning cover a continuous range of different time scales, with long-, medium-, and short-term growth variations being controlled by very different internal factors influenced by the environment. Misson et al. ascribed what they called long- and medium-term radial growth variation of individual trees mainly to structural adaptation, such as enlargement of the crown or the root system. Short-term radial growth variations, on the other hand, were considered to be due mainly to physiological acclimatisation, with response to factors such as stomatal conductance regulation, variation of photosynthetic capacity, and respiration. Nevertheless, it is clear that the adaptations are linked, and that radial growth is the integrative response of this mutual inter-relationship.

SELECTION CUTTING

The results of selective cutting of Ponderosa Pine.

Selection cutting, also known as selective cutting, is the silvicultural practice of harvesting trees in a way that moves a forest stand towards an uneven-aged or all-aged condition, or 'structure'. Using stocking models derived from the study of old growth forests, '"Selection cutting"', also known as 'selection system', or 'selection silviculture', manages the establishment, continued growth and final harvest of multiple age classes (usually three, but 5 or even 10 are possible) of trees within a stand. A closely related approach to forest management is Continuous Cover Forestry (CCF), which makes use of selection systems to achieve a permanently irregular stand structure.

Selection cutting or systems are generally considered to be more challenging to implement and maintain than even-aged management, due to the difficulty of managing multiple age classes in a shared space, but there are significant ecological benefits associated with it. Uneven-aged stands generally exhibit higher levels of vertical structure (key for many species of birds and mammals), have higher levels of carbon sequestration, and produce a more constant flow of market and non-market forest resources than even-aged stands. Although a forest composed of many stands with varied maturity ages may-be comparable, this would be at the forest rather than the stand level. This silvicultural method also protects forest soils from the adverse effects of many types of even-aged silviculture, including nutrient loss, erosion and soil compaction and the rapid loss of organic material from a forested system. Selection silviculture is especially adept at regenerating shade-tolerant species of trees (those able to function under conditions of low solar energy, both cooler and less light), but can also be modified to suit the regeneration and growth of intolerant and mid-tolerant species. This is one of many different ways of harvesting trees. Selection cutting as a silvicultural system can be modified in many ways and would be so done be a forester to take into account varied ownership goals, local site conditions and the species mix found from past forest conditions.

Selection cutting is often (sometimes deliberately) confused with "selective" cutting, a term synonymous with the practice of highgrading (the removal of the most economically profitable trees in a forest, often with a disregard for the future of the residual stand). Often the latter term is used by foresters or loggers to imply the former (which has a generally positive connotation in forestry circles) and mislead landowners into stripping their woodlot of its most valuable timber. Used correctly, the term 'selection cutting', 'selection system', or 'selection silviculture' implies the implementation of specific silvicultural techniques—usually either 'single tree selection', 'group selection' or a combination of the two—to create an uneven-aged or all-aged condition in a forest stand, one more akin to a late successional or 'climax' condition.

Partly as a result of such confusion, the term Plenterwald, which is the German term for selection cutting, is being more commonly used as the standard term in English. Increasingly, especially in Britain, Ireland and elsewhere in Europe, the term Continuous Cover Forestry (CCF) has been adopted to embrace an approach to stand management that most often employs selection systems to achieve a permanently irregular stand structure.

Single-tree Selection

The most common type of selection system is single-tree selection, in which scattered individual trees of multiple age classes, whose canopies are not touching, are harvested. This type of selection system generally produces small canopy openings especially conducive to the establishment and growth of shade tolerant tree species.

Group Selection

Another variation of selection silviculture is group selection. Under this system, a number of 'groups', or small openings created by the removal of several adjacent trees, are created in complement to the harvest of scattered individual trees. If the groups created are large enough, and if seed-bed conditions are favorable, this can allow species which are intolerant of shade to regenerate. Group selection is designed to mimic larger, multi-tree mortality events, which in some environments may represent natural disturbance regimes.

The maximum size of a group (before it becomes a patch, or clearcut) is debatable. Some say it may be up to 2 acres (0.8 hectares) in size, whereas others limit it to a maximum of 0.5 acres (0.1 hectares).

In any case Plenterwald can operate in a small areas of 1/3 - 1/2 hectare, whereas other systems need a bigger area. Behind this is the philosophical idea that a stand should be balanced (that is equal amounts of land cover for each age class) in the same way that a forest would be balanced under a clear cut régime (that is stands collectively are balanced on yield flow). The reasoning is based on the Normalwald concept, which is a model of a forest over 100 years that will produce an amount of money that is consistent over time with treatments being consistent over time rather than big expenses or big profits at one time and low expenses and low profits at another.

Care need to be taken to avoid epicormic shoots growing on trunks of surrounding trees such that they lead to knotty wood, if timber production is desired. It is also challenging to visualize the groups with cuttings over time.

Implementing a Selection System

In North America, trees are selected for harvest in a selection system with reference to the Arbogast Method (named after the method's creator). This is also known as the BDq method. Under this method, a harvest is specified by defining a residual basal area (B), a maximum diameter (D), and a q-ratio (q). The q-ratio is the ratio of the number of trees in a diameter class to the number of trees in the next larger class. Typically diameter classes are either 4 centimeters or 2 inches.

When the Q is plotted on semi-log paper it gives a straight slope for uneven aged stands. However, in reality this slope can be seen to vary from what is called an S-curve in old growth forests to cut off the older trees giving a reverse-J curve in a managed stand. The curve is also an ideal curve and there may be variations to some extent, particularly in earlier number of trees where there are many more seedlings and saplings than the model Q-ratio would suggest.

Given the BDq, a curve representing the state of the residual stand is computed. This curve is compared to the inventory data from a stand, specifically the curve of the diameter classes of the trees in the stand against the number of trees in each diameter (age) class. Diameter is used as a surrogate for age and thus called an age class even though strictly it should be a size class. The comparison of these two curves tells the forester how many trees of each age-class should remain in the stand. Surplus trees are marked for harvest. If there are too few trees in a class, the forester will determine if it is necessary to reduce the removal of trees from neighboring classes to maintain an ideal q-ratio.

The goal of the use of a BDq curve is to ensure the continued development of trees in each age class, and the continued availability of mature timber to harvest on a relatively short cutting cycle (8–15 years). Longer cutting cycles may be used depending on species mix, silvicultural goals and if the aim is amenity or economic forestry in respect to the land.

Following this method with well performed forest inventories should see the right amount of cutting. However, reality has shown about a third of forests are overcut and a third are undercut. It appears that the model also departs from reality in many cases, and so cannot be solely relied on. The judgement of an experienced forester is also needed.

VARIABLE RETENTION

Variable retention is a relatively new silvicultural system that retains forest structural elements for at least one rotation in order to preserve environmental values associated with structurally complex forests.

Some examples for environmental values are forest cover connectivity, soil stabilization, microclimate under retained trees, habitats associated with live or dead trees and species diversity due to habitat preservation, and wildlife corridor preservation. Variable retention also aids in emulation of natural disturbances by leaving behind some residual structure from previous stand which is typical for stand replacing disturbances Traditional silviculture systems such as clearcut, patch cut, shelterwood, etc. are focused on maximizing timber production and future regeneration of the trees. Variable retention on the other hand is focused on what is retained.

Value of Variable Retention

Variable retention is gradually becoming a popularly promoted tool for helping resolve the dilemma between demands for wood and demands to maintain habitat and ecology biodiversity and structural diversity in managed forests. What is at issue is how much a forest operation can log without adversely interfering with other ecological processes within the forest.

Variable retention minimizes the impact of logging operation by leaving biological legacies such as coarse woody debris(nurse logs and snags). Either few trees or many trees can be retained under the variable retention system, and trees can be retained in patches (aggregated retention) or left uniformly throughout a stand (dispersed retention); hence the name "variable retention." It is a technique for retaining trees as key structural elements of a harvested stand for at least until next harvest rotation in an effort to maintain species, habitat diversity and forest-related processes. A 2014 meta-analysis of retention forestry research found "support for wider use of retention forestry since it moderates negative harvesting impacts on biodiversity."

There are four key mechanisms through which variable retention is presumed to maintain biodiversity:

- By providing a constant supply of structural features that are at high risk to being lost due to modern forestry practices and that are known to be important to habitat availability, such as large trees, very young trees, snags, and coarse woody debris.
- By providing adequate refuge for sensitive species that will colonize the surrounding managed forest environment as it develops suitable conditions.
- By establishing habitat patches, Patch dynamics, that can serve as stepping stones for the dispersal of newly produced offspring, seeds, and spores.
- By increasing the structural diversity of managed stands.

Variable Retention - Harvesting Methods

The method of variable retention is a more deliberate process than clearcutting. In some cases helicopters take the logs out. The use of the helicopters avoid the need to build logging roads inside the forest and hence avoid the disturbance of the forest floor. Yet, helicopter logging is less fuel efficient than truck use. Some logging operations that use variable retention try to emulate forest disturbances. In those cases, deadwood in snags and logs is left behind. Such dead or dying trees provide a growth medium for fungi and epiphytes, as well as food and shelter for insects and the birds. Loggers also leave younger trees that will continue to grow.

Variable retention harvesting retains more than 15% of the original stand in both rolling and permanent pockets of untouched trees and refugia patches. Retained forest structure is composed of live and dead trees that enhance structural diversity, and provide a hospitable habitat for variety of living organisms that prevail in the unlogged forest. This silviculture regime provides post-harvest ecological structure while creating sufficient opportunity to plant and naturally regenerate valuable tree species for timber, as well as restore historical coniferous and/or deciduous tree dominance to the forestland. Some timber companies have restricted the use of variable retention silviculture to only poorly stocked stands of rare but valuable tree species.

Controversy

Ecologists regard the variable retention method of harvesting as being far less detrimental to the forest ecosystem than clearcutting. However, opponents claim that variable retention is much more laborious, tedious, time-consuming and expensive than clearcutting. A review of studies in British Columbia and in similar forest types indicates that while both aggregated and dispersed retention can contribute to biodiversity conservation there may be some negative consequences for timber production such as wind damage to retained trees and reduced growth rates of tree regeneration compared to clearcuts, concluding in favour of an adaptive management approach for balancing competing objectives of production of wood and biodiversity conservation.

Forestry organizations, such as the Forest Stewardship Council (FSC), approve of variable retention and are now declaring timber obtained through such a harvesting method as certified wood for green building. Many forest operations in the Canadian province of British Columbia, including MacMillan Bloedel, use the variable retention method to harvest timber in the temperate rain forests along the Pacific coast.

Many logging companies have not maintained a total commitment to phasing out clearcutting and embracing variable retention harvesting, and even MacMillan Bloedel had occasionally fallen short of its commitment to phase out clearcutting. Although world markets are beginning to ask for certified wood obtained entirely through variable retention harvesting, it is not yet clear whether the general population of consumers will be willing to pay higher prices for such wood.

SEED TREE METHOD

An even aged regeneration method in which a new age class develops from seeds that germinate in fully exposed microenvironments after removal of all the previous stand except a small number of trees left to provide seed. Seed trees are removed after regeneration is established.

A slash pine stand, with half the original stand still intact (right side) and the other half following the seed-tree cut (left side). Once the new cohort is established, the seed-trees will be removed.

A loblolly pine stand showing both the seed trees and the newly established cohort. The seed-trees will be removed after the new cohort is allowed to grow for one or two more years.

The seed-tree regen method is very common where natural regeneration is desired and the species to be regenerated are shade-intolerant and light-seeded.

Cuts Involved

The most common application of the seed-tree regen method involves two cuts.

- The first is often referred to as the seed-tree cut. Its intent is to remove the majority of the trees in the stand to create a fully exposed microclimate for the establishment of seeds from a small number of high-quality mature trees left on site for this purpose. With the exception of leaving the seed-trees, this operation is very similar to a clearcut.
- The second cut is often referred to as the final or removal cut, and occurs between 3 and 10 years following the seed-tree cut, depending on the species being regenerated. Its intent is to remove the seed-trees to release the newly established cohort. There may be significant value left in the seed-trees that can be captured with this harvest. This cut is necessary even if the seed-trees are not merchantable because they are too large for the mills, have been damaged by wind or lightning, or are simply not valuable enough to be worth paying a logger to set up on the site. If the seed-trees are not removed, they will suppress the growth of the new cohort.

The seed-tree regen method results in an even aged, single cohort stand upon completion of the removal cut. For a brief time between the seed-tree and removal cuts, the stand is two aged, with both the seed-tree cohort and the newly established cohort coexisting. However, because this is only a temporary condition, and the intent is the regeneration of an even aged stand, the seed-tree system is classified as even aged.

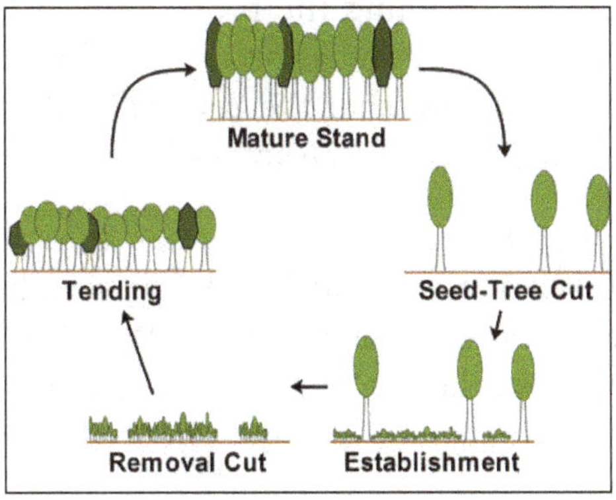

A loblolly pine stand immediately prior to the removal cut.

Common Variations

The two most common even aged variants of the seed-tree regeneration method are the uniform seed-tree and the aggregated seed-tree. Both include the same cuts, but differ in the spatial arrangement of the seed trees.

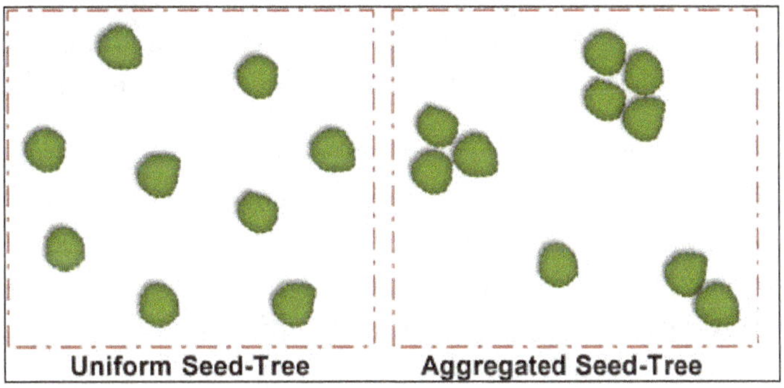

Overhead view of uniform versus aggregated seed-tree distributions.

Uniform spacing is more common as it more effectively distributes seed across the stand with a minimum number of seed trees.

However, leaving aggregated patches of seed trees does offer several advantages:

- Aggregated spacing may limit damage to the seed trees during the seed-tree cut.
- If seed trees are susceptible to windthrow or lightning strikes, leaving multiple trees at each location will lessen the probability that a disturbance removes all seed trees from that area. Even if one or more seed trees are lost, there are several at each location, reducing the risk of a regeneration failure.
- These small patches of large trees may also offer improved habitat for some wildlife as compared with uniform spacing.
- Aggregated spacing can also improve operability during the removal cut and

may minimize damage to the new cohort by lessening the proportion of the stand affected by skidding. If it is necessary to leave more seed trees than required to regenerate the stand primarily to ensure that the removal cut is a merchantable harvest, then leaving them in groups may be advised.

- Seed trees may also be retained in strips. This is more efficient for harvest operations. The strips can be placed so that they run perpendicular to the prevailing wind direction to maximize seed dispersal distances over cut areas, although the probability of windthrow should be considered before prescribing this. Strips can also be located along contours or up and down steep slopes, depending on the harvesting system being utilized.

Ecological Considerations (Silvics)

The seed-tree regen method creates a fully exposed microclimate for the establishment of the new cohort. Thus, it is only suitable for species that are intermediate, intolerant, or very intolerant of shade.

Seed must naturally disperse from the retained seed-trees. Therefore this method is best suited to species that are light-seeded, particularly those with wind-dispersed seeds. Heavy seeded species or species with seeds that are subject to high predation such as oaks or hickories are not well suited to the seed-tree regen method.

Retained seed-trees must remain on site for 3 to 10 growing seasons for most species to ensure adequate regeneration. It is important that seed-trees remain healthy and vigorous during this period to continue producing seed, and so they are merchantable at the removal harvest. Due to these constraints, species that are deep-rooted, and as a result relatively wind-resistant, are best suited to this method of regeneration.

FOREST NURSERY

Tree seeds germinate and trees grow readily under natural conditions. These delicate young seedlings and young trees are exposed to adverse dry season conditions and intense competition from other plants in natural forests or plantations. Forest nurseries can provide optimum care and attention to seedlings during their critical juvenile stage, resulting in the production of healthy, vigorous seedlings. In many cases, successful reforestation requires nursery grown seedlings, since degraded areas have un-favourable conditions making natural vegetation or direct seeding not feasible.

Forest nursery is a managed area, designated to produce tree seedlings grown under favourable conditions until they are ready for planting. It can be an informal, small-scale arrangement or a large commercial enterprise. Nurseries vary in size, facilities (supplies, tools, equipments, etc.), types of seedlings produced and nursery operations. They also differ significantly in quality and quantity of planting stocks produced.

However, all nurseries primarily aim to produce sufficient quantities of high quality planting stocks to satisfy the needs of seedling users. Users include the nursery owners themselves, individuals, community organizations, farmer groups, government agencies, non-government organizations, corporate or private customers.

Nurseries may often provide income generating opportunities for the operators and enhance social capital and technical capacity and leadership skills of communities. Smallholder tree nurseries also serve as important training and research ground for many smallholder farmers.

Forest nurseries are defined in a numerous way:

- An area or place where forest seedlings are grown.
- An area where tree seedlings and plants are raised for eventual planting out.
- An area where trees, shrubs, or plants are grown for transplanting, for use as stocks for budding and grafting.
- A production unit that grows planting stock (seedlings and saplings) of forest trees and shrubs.
- A plot of land where young trees can be raised under more or less controlled conditions.

Forestry projects and large scale afforestation and reforestation programmes spawn the establishment of self-help and self-sustaining local nurseries, which remain functional even after project termination. These local nurseries may be linked in a network to meet project, community or market needs. Government and non-government agencies often provide modest support to enhance the development of local nurseries, in recognition of the spontaneous growth of local nurseries.

These efforts strengthen local livelihood development and sustain public reforestation initiatives. In many areas, where nursery development has flourished, farmers also developed keen interests in collecting and propagating diverse species, including indigenous trees. Such forest nurseries have become rich repositories of tree biodiversity.

Significance of Forest Nursery

Good nursery operations incorporate the selection of the best quality seeds and seedlings, thus initiating tree improvement. Forest nurseries are advantageous in propagating recalcitrant seeds, dormant seeds and tree species with irregular flowering. Nurseries are also good venues for vegetative propagation operations. Some important species do not need seed every year.

Plantation of these species can be raised annually only by collecting all available seed in years and sowing it in nursery to raise seedlings to be planted out in various years. Some species grow very slowly and if the seed of these species is sown directly in plantation, the seedlings are most likely to be swamped by weeds and killed. Therefore, slow growing species are generally raised in nursery and planted only when the seedlings are not liable to be damaged by weeds.

Success of roadside avenue plantation depends largely on planting tall and sturdy plants which can only be obtained from nursery. Plantations of some species when by direct sowing are not so successful as when raised by planting. In such cases, nursery is an essential part of artificial regeneration of those species.

The best method of introduction of exotics, viz. tropical pines, poplars, eucalyptus, etc. is only by planting and therefore nursery is very essential for them. Planting of nursery-grown plants is surest method of artificially regenerating poor and barren sites. Causalities in plantations have to be replaced either in the year of formation or in the next year.

Sowing done in the gaps are liable to be unsuccessful as a result of suppression from weeds and cannot catch up the plants from original sowing. Therefore, replacement of causalities is always done by planting nursery-grown plants or stumps and so nursery is very essential for causality replacement.

Classification of Forest Nurseries

- On the basis of area size:
    - Small Nursery: Nurseries with an area of less than 3 ha.
    - Medium Nursery: Nurseries with an area of 3-10 ha.
    - Large Nursery: Nurseries with an area of more than 10 ha.
- On the basis of irrigation facility:
    - Dry Nursery: It is a nursery that is maintained without any irrigation or other artificial watering.
    - Wet Nursery: It is a nursery that is maintained by irrigation or other artificial watering during the dry periods.
- On the basis of planting stock:
    - Bareroot Nursery: In bareroot nursery, plants grow directly in the nursery soil and the roots are separated from the soil at the time of lifting. The lifted planting stock is further handled and planted without soil surrounding the roots.

    - Container Nursery: In container nursery, plants are grown with roots in a growing medium held in a container. Roots of lifted plants are covered by the adhering growing medium.

- On the basis of size of seedlings:
    - Seedling Nursery: A nursery which has only seedling beds i.e., in which seedlings only are raised, no transplanting being done is called seedling nursery.
    - Transplant Nursery: A nursery which has only transplant beds, in which seedlings are transplanted for preparation for forest planting, is called transplant nursery. In general, separate seedling and transplant nurseries are seldom made.

- On the basis of ownership:
    - Institutional Nurseries: They are commonly established, operated and maintained by research institutions, academic institutions, corporations and international organizations. They are permanent in nature but may be larger, medium or small, based on purpose.
    - Government Nurseries: They are established and operated by national or local government to support reforestation and community tree planting programmes. They are large in size and are permanent in nature.
    - Industrial Nurseries: They are commonly established and operated by large integrated wood based industries. They are large in size with high production capacities and permanent in nature. Seedling production complies with rigid quality standards or specifications.
    - Project Nurseries: They are established and managed by projects and non-governmental organizations (NGOs) or development organizations to promote tree planting culture within target communities. The intended lifespan of project nurseries is usually 3-5 years, but may continue for longer periods.
    - Community Nurseries: They are established by communities to support tree planting programmes for social benefits.
    - Group Nurseries: They are also established to build technical and leadership capacities of group members and expand the number of species and quality of germplasm available to group members.
    - Individual Nurseries: They are also known as household nurseries or backyard nurseries and are generally established and managed by individual farmer or household. These nurseries are usually initiated to produce seedlings for planting on operator's personal farms. The house compound or farmland provides the planting sites. They are normally small in size.

- On the basis of duration of the use:
    - Temporary Nursery: It is a nursery that is maintained for supplying stock for a short period after which it is abandoned. Normally, it is constructed in the plantation area and usually small in size. It is suitable for hilly regions.
    - Permanent Nursery: It is a nursery that is maintained for supplying nursery plants for a long time on a permanent basis. It is intended to meet the requirements of one or more ranges and it is relatively larger in extent.

Characteristics of Temporary Nursery

Features

Temporary nurseries are constructed for a short period of time and smaller in size. Manuring is not necessary and mostly located near/inside the planting area which is appropriate chiefly for casualty replacement. Elaborate soil preparation is not necessary and cost of transportation of seedlings to the planting sites is low. Seedlings of different species for mixed crops are raised. Gap between lifting the stock from the nursery and actual planting is less.

Advantages:

Temporary nurseries are usually constructed in newly cleared sites fairly rich in humus and so, manuring is not required. Eucalyptus nurseries are an exception to the general rules. Minimum trouble with the weeds, destructive insects and diseases are observed. It enables raising of species in their optimum altitudinal zone in hills. Cheap transport of planting stock without any serious damage or shock.

Disadvantages:

They are comparatively costly and difficult to supervise. Proper supervision is not possible as it is made out of way places. Due to lack of irrigation facility, the growth of seedlings is usually slow and heavy mortality is observed.

Characteristics of Permanent Nursery

Features

Permanent nurseries are fit for large and intensive work and intensively managed areas. It is established where all the facilities are available i.e., easy supervision, communication facilities, labours, etc. Intensive manuring and soil working are done in perpetuity.

It is used for large scale afforestation works, or distribution to the villagers under community and private forestry programme. Large labour forces, tools and equipments are needed. Initial cost of formation is high but is cheaper in the long run. Regular skilled supervision is mandatory.

Advantages:

It supplies varieties of planting stocks such as root-shoot cuttings, grafted plants, layering, budding, container seedlings, etc. Duration of service life is long and production cost is reasonable. It meets the requirement of larger areas/ranges/divisions. Supervision cost is low and can be easily supervised. Plants are raised year after year for a long time on same site.

Disadvantages:

Transportation of seedlings is difficult and costlier and establishment cost is high. Manuring of beds annually and intensive soil working are essential. It requires large labour forces throughout the year which is difficult during agricultural seasons.

Seedling Quality

The end user and the environmental conditions at the site where the seedlings will be planted determine what kind of seedling is needed. A researcher will require planning stock raised under uniform conditions to ensure that the research results are not confounded by the status of the seedlings. On the other hand, farmers will require seedlings that can survive under hard, dry and nutrient-deficient conditions.

The good looking seedling at the nursery is worthless if it does not survive and grow after planting out. Targeting seedling production to the anticipated field site is important in producing strong healthy seedlings.

Seedlings need to be hardened off and regularly prepared for real-life harsh conditions by withholding water from time to time after they have developed a strong and healthy root system. Insufficient care in the nursery leads to weak and retarded plants with small root systems that will not survive in a harsh environment.

Quality seedlings produced for different sites may look different from each other but they all have one thing in common: a well-developed root system with many root tips from which new roots can quickly develop. In areas with adverse environments such as dry, flooded, saline or nutrient-deficient sites, only well-developed plants have a good chance of survival. Seedlings with deeper root system are preferred for dry areas and larger plants are preferred for weedy sites so that they can outgrow weeds quickly. Many studies have proved that field survival and productivity are related to the quality of the seedling used.

Seedling quality depends on the following parameters:

- The ability of plant to produce new roots quickly.
- The speed with which seedlings get anchored in the ground and start assimilating and growing after planting out.

- A well-developed root system
- Sun-adapted foliage.
- A large root collar diameter.
- A high root: shoot ratio.
- Good carbohydrate reserves.
- An optimum mineral nutrition content.
- The establishment of adequate mycorrhizal or Rhizobium infection.

Nursery Planning and Management

Quality seedling production depends on a sound understanding and implementation of the key principles of nursery planning and management. As with the operation of any forest nursery, planning is essential for a successful and productive nursery. Factors to be taken into account are efficient use of inputs and resources, identifying problems and potential solutions, monitoring progress and working towards nursery objectives.

Nursery plans vary in detail and in implementation. Nursery plan may be formal written documents or a list of informal management guidelines. Generally, project and group nurseries engage in participatory planning, working with stakeholders and members.

Recommended elements of a nursery plan include: site selection, objectives, demand forecasting and target seedling production, inputs and scheduling, contingency plans to respond to unexpected problems, documentation, staff training, promotion and marketing and networking.

Nursery Site Selection

The selection of an appropriate nursery site is the most important decision affecting the efficient production of good quality plants.

The following are some of the technical factors which need to be considered while setting a nursery:

- Water Supply: In order to have the good irrigation facilities, it should be situated near or slightly below the source of adequate water supply. Reliable and adequate water supply is essential. Water source should be near and at the higher level than nursery site. Water should be available throughout the year.
- Availability of Suitable Soil: For raising seedlings in polythene bag (tube), soil of the actual nursery itself is not important. There should be suitable source of

soil preferably forest topsoil and sand within easy reach of the nursery. Clayey soils should not be selected as their aeration and drainage is poor and they are liable to crack during summer.

- The symbiotic organism (Mycorrhiza and Rhizobium) should be present in soil because most of the trees grow best if their roots are associated with certain symbionts, which help in their nutrition. Mostly these organisms are found naturally but in some cases artificial inoculation are needed.
- Access: Nursery should be situated as centrally as possible with reference to the area to be planted. It should preferably be near residence area for close and regular supervision. Nursery should be near to the plantation site to facilitate easy transportation of seedlings and access road should be usable at all seasons of year.
- Aspect: Slope with South aspect is much warmer and chosen for hill nursery at high elevation while in low elevation north facing slope is chosen.
- Slope: The ideal slope is about 5 degree which is steep enough to allow proper drainage. Complete flat land should be avoided; it is likely to become water logged during monsoon rains. Steeper slopes will have to be terraced and very steep slopes should be avoided as it may be difficult to make a nursery bed and path on each side to allow access to the beds.
- Exposure to Frost, Strong Winds and Flooding: At high altitudes, sites which are particularly liable to frost damage should be avoided. Sites exposed to strong winds and with danger of flooding or landslides should be avoided.
- Labour Availability: Labour should be available without difficulty preferably near a village so that labourers need not walk much distance for work.
- Availability of Land: There should be enough land to raise the number of seedlings needed and if possible room for expansion. Legal enquiries should be made about the legal ownership of the land.

Rooted Cuttings in Forest Nursery

Cuttings rooted in a nursery are commonly set in raised, well-drained beds of friable soil. Before planting cuttings, the lower end may be cut again to ensure that the cut surface is clean. This cut at the base is made just below a node, because generally root formation is best at the nodes. If the bark is thick, it may be sliced or scraped off the base of the cutting to make it easier for new roots to break through.

Push the cuttings into the soil until about 2/3 of the length of the cutting is covered. For cuttings of trees, two buds are allowed above-ground. If they both grow out, the weaker one is removed so that the most vigorous one will form the tree trunk. In the case of shrubs 3-5 buds may be left aboveground and all emerging shoots are allowed to grow. The cuttings may be set at an angle.

A vertically placed cutting tends to produce one dominant shoot, whereas a more horizontally inclined cutting is likely to produce several more or less equivalent shoots. Hence a vertical position is generally preferred for trees and a slanted position for shrubs. The cuttings should not be planted upside down. The cuttings are planted in rows about 25 cm apart with a spacing of about 20 cm in the row. After planting, the bed is watered again.

Cuttings can also be set in pots, but this requires close attention to watering. The soil must be kept moist all the time, but if it becomes too wet the young roots quickly die because of lack of oxygen. In a bed, it is easier to keep the soil moisture between these limits. Shade helps to reduce moisture loss and is essential for cuttings that retain some leaves, also because shade reduces air temperature.

Shelter is also important because it protects the cuttings against drying wind and tends to raise the humidity of the air. Whereas seedlings have a dominant taproot, cuttings form so-called adventitious roots which are basically all equal. In due course some adventitious roots may become much stronger than the others, but the root system does not become as strong as that of a seedling with a taproot.

Stumps

They are root shoot cuttings of seedlings. Cutting root and shoot is known as stump. Plants with a collar diameter of 1.5-2.0 cm and taproot with a length of not less than 30 cm are selected. Shoots are cut off at 3 cm and all the lateral roots are pruned.

The tap roots should be cut at a distance of 20-30 cm from the collage region. It is most common for propagating teak (Tectona grandis) and gamhar (Gmelina arborea). They are easily planted in the field and are very easy for handling and transportation.

Plant Propagation Structures and Forest Nursery

The need for enclosed structures varies greatly with the climate. In humid locations, structures may be needed only to provide shade or to keep pests out, whereas under arid conditions enclosures are needed to maintain higher humidity for germinating seed, rooting cuttings and successful grafting.

Although automated misting systems may be desirable, simple structures enclosed in polyethylene, such as polypropagator or polytunnel are low cost and sufficient in most cases where small number of plants are produced.

Some of the important plant propagation structures are presented below:

Greenhouse/Glasshouse

A greenhouse (also called glasshouse) is a propagation structure in which plants are grown. These structures vary in size based on the purpose and size of plants grown.

Greenhouses allow for greater control over the growing environment of plants. Depending upon the level of greenhouse, key factors which may be controlled include temperature, light, water, fertilizer and atmosphere. Greenhouses may be used to overcome the poor growing qualities of a piece of land, such as poor light levels, seasons, or soil infertility. Greenhouses allow certain plants to be grown throughout the year.

A greenhouse is a structure with different types of covering materials, such as a glass, fibreglass, double sheets of polyethylene (PE) film or plastic roof and frequently glass or plastic walls. Several new polycarbonate and acrylic materials are now made for greenhouse coverings. It heats up because incoming visible solar radiation (for which the glass is transparent) from the sun is absorbed by plants, soil and other things inside the building. Air warmed by the heat from hot interior surfaces is retained in the building by the roof and wall.

Mist Chambers

Nursery plants propagated by cuttings are grown in mist chambers and sometimes for germinating seeds. In mist chamber, relative humidity is maintained artificially at high level (95 per cent) with the help of mist installations, which spray water under pressure. Artificial microclimate and high relative humidity facilitates better root initiation and development whereas cooling effect prevents the cutting from drying out.

Higher success rate are found in propagation of hard wood cuttings under mist chamber conditions. Nowadays, mists chamber with fully automatic system and timers are installed which maintains temperature and humidity even in unfavourable environmental conditions.

Net Houses

Net houses are widely used as propagation structure in tropical areas, where artificial heating is not required and artificial cooling is expensive. The roof of net house may be covered with gunny cloth or even with live plant creeper to cut off the solar radiation and to keep the house cool. Different types of shade nets are available in the market for shade purpose. Net houses can be constructed as per need of propagation and therefore size of net houses varies.

Poly-houses

Poly-house is widely used for propagating off season seedlings which fetch better price in the market. Modern poly-houses are automated for temperature and humidity control to assist propagation. In poly-houses, plants are propagated using pro-trays or poly-bags on ground or benches. In Poly-houses, thermostat, hygrometer, automatic ventilation system is maintained to facilitate microclimate. Irrigation and fertigation systems in poly-houses are automated to provide measured quantity of water and nutrients to plants.

Plastic Tunnels

Plastic tunnel is a simple but effective method of protective cultivation, used by many nurserymen throughout the world. The three feet wide loops which support the polyethylene are made from 0.2 inch diameter wire and are erected at 30 inches intervals. A white translucent polyethylene sheet of 6 feet wide is then stretched over the loops and is fastened to the frame of the tunnel. Plastic tunnels prove effective for seed germination as well as vegetative propagation of nursery plants in winter season.

Walk-in Poly Tunnels

It is constructed for larger operations or larger plants. Walk-in structures like tunnel is made from semi-circular steel rods and covered with polyethylene sheeting.

Propagation Boxes

For small operations and small plants, propagation boxes are sufficient. These boxes can be made to varying specifications according to the purpose. A recommended size is 2 m long, 1 m wide and 0.5-1.5 m high, with a slanting lid. Layers of stones, gravel and rooting substrate are arranged inside the box. The box is completely enclosed in polyethylene, i.e. including the bottom, depends on the conditions and the main use. For propagating cuttings, a complete enclosure is recommended to guarantee very high air humidity.

References

- Nyland, Ralph D. (2002). Silviculture: Concepts and Applications. McGraw-Hill series in forest resources (2nd ed.). Boston: McGraw-Hill. p. 20. ISBN 978-0-07-366190-2
- Helliwell R., Wilson E. R. (2012). "Continuous cover forestry in Britain: challenges and opportunities". Quarterly Journal of Forestry. 106 (3): 214–224
- Silviculture-textbook-sp-9418, faculty, Stovall, silviculture, seed-tree-159: forestry.sfasu.edu, Retrieved 16 April, 2020
- Pond, Nan C.; Froese, Robert E. (2015). "Interpreting Stand Structure through Diameter Distributions". Forest Science. 61 (3): 429–437. doi:10.5849/forsci.14-056
- Forest-nursery, forest-nursery-meaning-significance-and-classification-forestry-4816: environmentalpollution.in, Retrieved 30 June, 2020
- Smith, Nicholas J.; Beese, William J. (2012). "Effects of Low Levels of Dispersed Retention on the Growth and Survival of Young, Planted Douglas-Fir". Forests. 3 (2): 230–243. doi:10.3390/f3020230

Sustainable Forest Management

CHAPTER 4

Forest management deals with the economic, social, technical, environmental and ecological aspects of forest. Sustainable forest management maintains balance between all these aspects by following the principles of sustainable management. This chapter discusses the subject of sustainable forest management in detail.

Sustainable forestry is widely discussed and almost universally desired, but tangible standards, goals, targets, or thresholds to evaluate sustainability are often defined vaguely. Identifying such standards is complicated by the diversity of forests, the diversity of objectives among forest decisionmakers (both public and private), the diversity of values among people who depend on forests, and the diversity of spatial and temporal scales that must be considered. The complexity of determining what is deemed to be sustainable, and by whom, leads to challenges in implementing and monitoring forest management. As a way of framing discussions about sustainability, we offer the following strategies which we believe are important considerations for sustainable forest management for ecoregions, States, and the Nation:

- Maintain a stable forest land base. This concept embraces strong sustainability and stresses the fact that goods and services provided by forests are irreplaceable. Thus, human-created capital is no substitute for forest land. Likewise loss of northern forests cannot be fully compensated by offsetting gains in forest area in other parts of the United States or the World.
- Maintain or increase forest biodiversity. This includes sustaining diverse populations of native plants and animals, diversity of forest ecosystems and habitats across the landscape, and genetic diversity of forestassociated plants and animals. Special emphasis is given to restoring species and habitats that are threatened with extinction, and controlling invasive species.
- Maintain diverse forest size structure and species composition on the landscape. Forests constantly change, but their trajectory of change is periodically altered by fire, insects, diseases, harvesting, weather, invasive plants, and other disturbances. Healthy, diverse forests can better cope with such disturbances and continue to function as forests.
- Maintain or increase the quality and quantity of water from forest ecosystems.

- Maintain or increase soil productivity and minimize soil erosion and contamination.
- Maintain or increase the capacity for sustained yield of timber and nontimber forest products and associated economic development.
- Maintain or increase forest-based employment and community stability. Commercial forest operations may be the most economical means of altering forest structure and composition in ways that are essential to achieving other goals such as habitat restoration, hazardous fuel reduction, or invasive species mitigation.
- Maintain or enhance the quantity and quality of forest recreation and other opportunities for people to experience forests.
- Maintain a system of institutions, policies, regulations, and incentives that support forest sustainability at multiple spatial scales.
- Increase environmental literacy and engage a wide range of stakeholders in sustainable forest management.

Many readers will note that these strategies cannot all be pursued equally, and pursuing some will require sacrificing the extent to which others can be pursued. Tradeoffs and compromises are inevitable in striving for sustainable forests. Sustainable forest management requires careful thought about the full range of benefits and consequences associated with management actions or inaction.

Sustainable forest management is complicated by the multiple benefits and the large temporal and spatial scales that must be considered. For example, short-term decisions about forest sector employment and harvest practices can have long-term effects on biodiversity. Similar tradeoffs can arise across large spatial scales; one State's effort to increase the area of forest reserved for wildlife habitat diversity could transfer timber harvesting to other States or to other nations and alter habitat diversity or sustained timber yield there. Urban and rural forests provide many of the same ecosystem services, but differ in the type and number of interactions with people.

Strong and Weak Sustainability

Sustainability can be further classified into weak and strong sustainability, with capital existing in three forms: Natural, Human and Human-made.

- Weak sustainability-Only concerned about the total capital value (natural + human + human-made) allowing for substitution among the three forms. Human-made or manufactured capital of equal value can take the place of natural capital lost as a result of development. As long as future generations are endowed with an amount of total capital that is not less than that of the current generation, the conditions of weak sustainable development are satisfied.

- Strong sustainability- Not merely concerned about keeping the aggregate stocks of capital constant, but also requiring maintenance and enhancement of natural capital levels (ecological assets and environmental quality) because the functions performed cannot be duplicated by manufactured capital.

Sustainable forest management is the management of forests according to the principles of sustainable development. Sustainable forest management has to keep the balance between three main pillars: ecological, economic and socio-cultural. Successfully achieving sustainable forest management will provide integrated benefits to all, ranging from safeguarding local livelihoods to protecting the biodiversity and ecosystems provided by forests, reducing rural poverty and mitigating some of the effects of climate change.

The "Forest Principles" adopted at The United Nations Conference on Environment and Development (UNCED) in Rio de Janeiro in 1992 captured the general international understanding of sustainable forest management at that time. A number of sets of criteria and indicators have since been developed to evaluate the achievement of SFM at the global, regional, country and management unit level. These were all attempts to codify and provide for independent assessment of the degree to which the broader objectives of sustainable forest management are being achieved in practice. In 2007, the United Nations General Assembly adopted the Non-Legally Binding Instrument on All Types of Forests. The instrument was the first of its kind, and reflected the strong international commitment to promote implementation of sustainable forest management through a new approach that brings all stakeholders together.

A definition of SFM was developed by the Ministerial Conference on the Protection of Forests in Europe and has since been adopted by the Food and Agriculture Organization (FAO). It defines sustainable forest management as:

> "The stewardship and use of forests and forest lands in a way, and at a rate, that maintains their biodiversity, productivity, regeneration capacity, vitality and their potential to fulfill, now and in the future, relevant ecological, economic and social functions, at local, national, and global levels, and that does not cause damage to other ecosystems."

In simpler terms, the concept can be described as the attainment of balance – balance between society's increasing demands for forest products and benefits, and the preservation of forest health and diversity. This balance is critical to the survival of forests, and to the prosperity of forest-dependent communities.

For forest managers, sustainably managing a particular forest tract means determining, in a tangible way, how to use it today to ensure similar benefits, health and productivity in the future. Forest managers must assess and integrate a wide array of sometimes conflicting factors – commercial and non-commercial values, environmental considerations, community needs, even global impact – to produce sound forest plans. In most cases, forest managers develop their forest plans in consultation with citizens, businesses, organizations and other interested parties in and around the forest tract being managed. The tools and visualization have been recently evolving for better management practices.

The Food and Agriculture Organization of the United Nations, at the request of Member States, developed and launched the Sustainable Forest Management Toolbox in 2014, an online collection of tools, best practices and examples of their application to support countries implementing sustainable forest management.

Because forests and societies are in constant flux, the desired outcome of sustainable forest management is not a fixed one. What constitutes a sustainably managed forest will change over time as values held by the public change.

Criteria and Indicators

Deforestation of native rain forest in Rio de Janeiro City for extraction of clay for civil engineering. An example of non sustainable forest management.

Criteria and indicators are tools which can be used to conceptualise, evaluate and implement sustainable forest management. Criteria define and characterize the essential elements, as well as a set of conditions or processes, by which sustainable forest management may be assessed. Periodically measured indicators reveal the direction of change with respect to each criterion.

Criteria and indicators of sustainable forest management are widely used and many countries produce national reports that assess their progress toward sustainable forest management. There are nine international and regional criteria and indicators initiatives, which collectively involve more than 150 countries. Three of the more advanced initiatives are those of the Working Group on Criteria and Indicators for the Conservation and Sustainable Management of Temperate and Boreal Forests (also called the Montréal Process), Forest Europe, and the International Tropical Timber Organization. Countries who are members of the same initiative usually agree to produce reports at the same time and using the same indicators. Within countries, at the management unit level, efforts have also been directed at developing local level criteria and indicators of sustainable forest management. The Center for International Forestry

Research, the International Model Forest Network and researchers at the University of British Columbia have developed a number of tools and techniques to help forest-dependent communities develop their own local level criteria and indicators. Criteria and Indicators also form the basis of third-party forest certification programs such as the Canadian Standards Association's Sustainable Forest Management Standards and the Sustainable Forestry Initiative Standard.

There appears to be growing international consensus on the key elements of sustainable forest management. Seven common thematic areas of sustainable forest management have emerged based on the criteria of the nine ongoing regional and international criteria and indicators initiatives. The seven thematic areas are:

- Extent of forest resources.
- Biological diversity.
- Forest health and vitality.
- Productive functions and forest resources.
- Protective functions of forest resources.
- Socio-economic functions.
- Legal, policy and institutional framework.

This consensus on common thematic areas (or criteria) effectively provides a common, implicit definition of sustainable forest management. The seven thematic areas were acknowledged by the international forest community at the fourth session of the United Nations Forum on Forests and the 16th session of the Committee on Forestry. These thematic areas have since been enshrined in the Non-Legally Binding Instrument on All Types of Forests as a reference framework for sustainable forest management to help achieve the purpose of the instrument.

On January 5, 2012, the Montréal Process, Forest Europe, the International Tropical Timber Organization, and the Food and Agriculture Organization of the United Nations, acknowledging the seven thematic areas, endorsed a joint statement of collaboration to improve global forest related data collection and reporting and avoiding the proliferation of monitoring requirements and associated reporting burdens.

Ecosystem Approach

The Ecosystem Approach has been prominent on the agenda of the Convention on Biological Diversity (CBD) since 1995. The CBD definition of the Ecosystem Approach and a set of principles for its application were developed at an expert meeting in Malawi in 1995, known as the Malawi Principles. The definition, 12 principles and 5 points of "operational guidance" were adopted by the fifth Conference of Parties (COP5) in 2000.

The CBD definition is as follows:

The ecosystem approach is a strategy for the integrated management of land, water and living resources that promotes conservation and sustainable use in an equitable way. Application of the ecosystem approach will help to reach a balance of the three objectives of the Convention. An ecosystem approach is based on the application of appropriate scientific methodologies focused on levels of biological organization, which encompasses the essential structures, processes, functions and interactions among organisms and their environment. It recognizes that humans, with their cultural diversity, are an integral component of many ecosystems.

Sustainable forest management was recognized by parties to the Convention on Biological Diversity in 2004 (Decision VII/11 of COP7) to be a concrete means of applying the Ecosystem Approach to forest ecosystems. The two concepts, sustainable forest management and the ecosystem approach, aim at promoting conservation and management practices which are environmentally, socially and economically sustainable, and which generate and maintain benefits for both present and future generations. In Europe, the MCPFE and the Council for the Pan-European Biological and Landscape Diversity Strategy (PEBLDS) jointly recognized sustainable forest management to be consistent with the Ecosystem Approach in 2006.

Independent Certification

Growing environmental awareness and consumer demand for more socially responsible businesses helped third-party forest certification emerge in the 1990s as a credible tool for communicating the environmental and social performance of forest operations.

There are many potential users of certification, including: forest managers, scientists, policy makers, investors, environmental advocates, business consumers of wood and paper, and individuals.

With third-party forest certification, an independent organization develops standards of good forest management, and independent auditors issue certificates to forest operations that comply with those standards. Forest certification verifies that forests are well-managed – as defined by a particular standard – and chain-of-custody certification tracks wood and paper products from the certified forest through processing to the point of sale.

This rise of certification led to the emergence of several different systems throughout the world. As a result, there is no single accepted forest management standard worldwide, and each system takes a somewhat different approach in defining standards for sustainable forest management.

In its 2009–2010 Forest Products Annual Market Review United Nations Economic Commission for Europe/Food and Agriculture Organization stated: "Over the years, many of the issues that previously divided the certification systems have become much less distinct. The largest certification systems now generally have the same structural programmatic requirements."

Third-party forest certification is an important tool for those seeking to ensure that the paper and wood products they purchase and use come from forests that are well-managed and legally harvested. Incorporating third-party certification into forest product procurement practices can be a centerpiece for comprehensive wood and paper policies that include factors such as the protection of sensitive forest values, thoughtful material selection and efficient use of products.

There are more than fifty certification standards worldwide, addressing the diversity of forest types and tenures. Globally, the two largest umbrella certification programs are:

- Programme for the Endorsement of Forest Certification (PEFC).
- Forest Stewardship Council (FSC).

The area of forest certified worldwide is growing slowly. PEFC is the world's largest forest certification system, with more than two-thirds of the total global certified area certified to its Sustainability Benchmarks.

In North America, there are three certification standards endorsed by PEFC – the Sustainable Forestry Initiative, the Canadian Standards Association›s Sustainable Forest Management Standard, and the American Tree Farm System. FSC has five standards in North America – one in the United States and four in Canada.

While certification is intended as a tool to enhance forest management practices throughout the world, to date most certified forestry operations are located in Europe and North America. A significant barrier for many forest managers in developing countries is that they lack the capacity to undergo a certification audit and maintain operations to a certification standard.

Forest Governance

Although a majority of forests continue to be owned formally by government, the effectiveness of forest governance is increasingly independent of formal ownership. Since neo-liberal ideology in the 1980s and the emanation of the climate change challenges, evidence that the state is failing to effectively manage environmental resources has emerged. Under neo-liberal regimes in the developing countries, the role of the state has diminished and the market forces have increasingly taken over the dominant socio-economic role. Though the critiques of neo-liberal policies have maintained that market forces are not only inappropriate for sustaining the environment, but are in fact a major cause of environmental destruction. Hardin›s tragedy of the commons has shown that the people cannot be left to do as they wish with land or environmental resources. Thus, decentralization of management offers an alternative solution to forest governance.

Countries participating in the UNREDD program and/or Forest Carbon Partnership Facility.

UN-REDD participants

Forest Carbon Partnership Facility participants

participants in both

The shifting of natural resource management responsibilities from central to state and local governments, where this is occurring, is usually a part of broader decentralization process. According to Rondinelli and Cheema, there are four distinct decentralization options: these are: (i) Privatization – the transfer of authority from the central government to non-governmental sectors otherwise known as market-based service provision, (ii) Delegation – centrally nominated local authority, (iii) Devolution – transfer of power to locally acceptable authority and (iv) Deconcentration – the redistribution of authority from the central government to field delegations of the central government. The major key to effective decentralization is increased broad-based participation in local-public decision making. In 2000, the World Bank report reveals that local government knows the needs and desires of their constituents better than the national government, while at the same time, it is easier to hold local leaders accountable. From the study of West African tropical forest, it is argued that the downwardly accountable and representative authorities with meaningful discretional powers are the basic institutional element of decentralization that should lead to efficiency, development and equity. This collaborates with the World Bank report in 2000 which says that decentralization should improve resource allocation, efficiency, accountability and equity "by linking the cost and benefit of local services more closely".

Many reasons point to the advocacy of decentralization of forest. (i) Integrated rural development projects often fail because they are top-down projects that did not take local people's needs and desires into account. (ii) National government sometimes have legal authority over vast forest areas that they cannot control, thus, many protected area projects result in increased biodiversity loss and greater social conflict. Within the sphere of forest management, as state earlier, the most effective option of decentralization is "devolution"-the transfer of power to locally accountable authority. However, apprehension about local governments is not unfounded. They are often short of resources, may be staffed by people with low education and are sometimes captured by local elites who promote clientelist relation rather than democratic participation.

Enters and Anderson point that the result of community-based projects intended to reverse the problems of past central approaches to conservation and development have also been discouraging.

Broadly speaking, the goal of forest conservation has historically not been met when, in contrast with land use changes; driven by demand for food, fuel and profit. It is necessary to recognize and advocate for better forest governance more strongly given the importance of forest in meeting basic human needs in the future and maintaining ecosystem and biodiversity as well as addressing climate change mitigation and adaptation goal. Such advocacy must be coupled with financial incentives for government of developing countries and greater governance role for local government, civil society, private sector and NGOs on behalf of the "communities".

National Forest Funds

The development of National Forest Funds is one way to address the issue of financing sustainable forest management. National forest funds (NFFs) are dedicated financing mechanisms managed by public institutions designed to support the conservation and sustainable use of forest resources. As of 2014, there are 70 NFFs operating globally.

Forest Genetic Resources

Appropriate use and long-term conservation of forest genetic resources (FGR) is a part of sustainable forest management. In particular when it comes to the adaptation of forests and forest management to climate change. Genetic diversity ensures that forest trees can survive, adapt and evolve under changing environmental conditions. Genetic diversity in forests also contributes to tree vitality and to the resilience towards pests and diseases. Furthermore, FGR has a crucial role in maintaining forest biological diversity at both species and ecosystem levels.

Selecting carefully the forest reproductive material with emphasis on getting a high genetic diversity rather than aiming at producing a uniform stand of trees, is essential for sustainable use of FGR. Considering the provenance is crucial as well. For example in relation to climate change, local material may not have the genetic diversity or phenotypic plasticity to guarantee good performance under changed conditions. A different population from further away, which may have experienced selection under conditions more like those forecast for the site to be reforested, might represent a more suitable seed source.

FOREST CONSERVATION

Forest conservation as the name suggests is the preservation and the protection of forests. It also involves the reversal of deforestation and environmental pollution. The preservation of all natural resources is absolutely essential for the balance of our ecosystem.

Importance of Forests

We are basically dependent on forests for our survival. And so their conservation is of essential importance:

- The most important function of forests is that it produces mass amounts of oxygen as a by-product of photosynthesis. Oxygen is the main respiratory gas for all animals, it ensures our survival.
- And while photosynthesis, trees also absorb carbon dioxide from the air. This is one of the main pollutants of air pollution. Hence forests also reduce air pollution.
- Forests also prevent soil erosion and keep soil pollution in check. Deforestation, in fact, leads to soil erosion on a large scale since the topsoil comes loose.
- Forests also play an important part in the water cycle and control moisture levels of our ecosystem.
- And finally, forests are the natural home and habitat for millions of species of animals, birds, and insects.

Ways to Conserve the Forest

- Controlled Deforestation: While deforestation cannot be avoided completely, we must look to control it. Young and immature trees should not be felled as far as possible. We must look to avoid large-scale commercial deforestation as well. Adapting practices such as clear-cutting or selective cutting will be beneficial in the long run.
- Protect against Forest Fires: Forest fires are the most common and deadly cause of loss of forests. They can start due to natural causes or can be accidents caused by man or even intentional in some cases. Once a fire spreads in a forest it is very

difficult to control. Precautions must be taken for such incidents. Making fire lanes, spreading chemicals to control fire, clearing out dry leaves and trees etc.

- Afforestation: This is the process by which we plant more trees in the area. We try to increase the forest cover by manual transplantation, or fresh plantation of trees. It is an attempt to balance our ecosystem to reduce the effects of deforestation and environmental pollutions of all types.
- Better Farming Practices: Slash and burn farming, overgrazing by cattle, shifting agriculture are all farming practices that are harmful to the environment and particularly to forests. We must keep all these practices under control.

REFORESTATION

A 21-year-old plantation of red pine in Southern Ontario.

'Reforestation' is the natural or intentional restocking of existing forests and woodlands (forestation) that have been depleted, usually through deforestation. Reforestation can be used to rectify or improve the quality of human life by soaking up pollution and dust from the air, rebuild natural habitats and ecosystems, mitigate global warming since forests facilitate biosequestration of atmospheric carbon dioxide, and harvest for resources, particularly timber, but also non-timber forest products. In the beginning of the 21 century more attention is given to the ability of reforestation to mitigate climate change as one of the best methods to do it.

Management

A debated issue in managed reforestation is whether or not the succeeding forest will have the same biodiversity as the original forest. If the forest is replaced with only one species of tree and all other vegetation is prevented from growing back, a monoculture forest similar to agricultural crops would be the result. However, most reforestation involves the planting of different selections of seedlings taken from the area, often of multiple species. Another important factor is the natural regeneration of a wide variety of plant and animal species that can occur on a clear cut. In some areas the suppression of forest fires for hundreds of years has resulted in large single aged and single species forest stands. The logging of small clear cuts and prescribed burning actually increases the biodiversity in these areas by creating a greater variety of tree stand ages and species.

Tropical tree nursery at Planeta Verde Reforestación S.A.'s plantation in Vichada Department, Colombia.

For Harvesting

A 15-year-old reforested plot of land.

Reforestation need not be only used for recovery of accidentally destroyed forests. In some countries, such as Finland, many of the forests are *managed* by the wood products and pulp and paper industry. In such an arrangement, like other crops, trees are planted to replace those that have been cut. In such circumstances, the industry can cut the trees in a way to allow easier reforestation. The wood products industry systematically replaces many of the trees it cuts, employing large numbers of summer workers for tree planting work. For example, in 2010, Weyerhaeuser reported planting 50 million seedlings. However replanting an old-growth forest with a plantation is not replacing the old with the same characteristics in the new.

In just 20 years, a teak plantation in Costa Rica can produce up to about 400 m^3 of wood per hectare. As the natural teak forests of Asia become more scarce or difficult to obtain, the prices commanded by plantation-grown teak grows higher every year. Other species such as mahogany grow more slowly than teak in Tropical America but are also extremely valuable. Faster growers include pine, eucalyptus, and *Gmelina*.

Reforestation, if several indigenous species are used, can provide other benefits in addition to financial returns, including restoration of the soil, rejuvenation of local flora and fauna, and the capturing and sequestering of 38 tons of carbon dioxide per hectare per year.

The reestablishment of forests is not just simple tree planting. Forests are made up of a community of species and they build dead organic matter into soils over time. A major tree-planting program could enhance the local climate and reduce the demands of burning large amounts of fossil fuels for cooling in the summer.

For Climate Change Mitigation

Forests are an important part of the global carbon cycle because trees and plants absorb carbon dioxide through photosynthesis. By removing this greenhouse gas from the air, forests function as terrestrial carbon sinks, meaning they store large amounts of carbon. At any time, forests account for as much as double the amount of carbon in the atmosphere. Forests remove around three billion tons of carbon every year. This amounts to about 30% of all carbon dioxide emissions from fossil fuels. Therefore, an increase in the overall forest cover around the world would mitigate global warming.

At the beginning of the twenty-first century interest reforestation grew over its potential to mitigate climate change. Even without displacing agriculture and cities, earth can sustain almost one billion hectares of new forests. This would remove 25% of carbon dioxide from the atmosphere and reduce its concentration to levels that existed in the early twentieth century. A temperature rise of 1.5 degrees would reduce the area suitable for forests by 20% by the year 2050, because it some tropical areas will become too hot. The countries that have the most forest-ready land are: Russia, Canada, Brazil, Australia, United States and China.

The four major strategies are: increase the amount of forested land through reforestation; increase density of existing forests at a stand and landscape scale; expand the use of forest products that sustainably replace fossil-fuel emissions; and reduce carbon emissions caused by deforestation and degradation.

Implementing the first strategy is supported by many organizations around the world. For example, in China, the Jane Goodall Institute, through their Shanghai Roots & Shoots division, launched the Million Tree Project in Kulun Qi, Inner Mongolia to plant one million trees. China used 2.4 billion hectares of new forest to offset 21% of Chinese fossil fuel emissions in 2000. In Java, Indonesia newlywed couples give whoever is conducting their wedding 5 seedlings. Each divorcing couple gives 25 seedlings to whoever divorces them. Costa Rica doubled its forest cover in 30 Years using its system of grants and other payments for environmental services, including compensation for landowners. These payments are funded through international donations and nationwide taxes.

The second strategy has to do with selecting species for tree-planting. In theory, planting any kind of tree to produce more forest cover would absorb more carbon dioxide from the atmosphere. However, a genetically modified variant might grow much faster than unmodified specimens. Some of these cultivars are under development. Such fast-growing trees would be planted for harvest and can absorb carbon dioxide faster than slower-growing trees.

Impacts on temperature are affected by the location of the forest. For example, reforestation in boreal or subarctic regions has less impact on climate. This is because it substitutes a high-albedo, snow-dominated region with a lower-albedo forest canopy. By contrast, tropical reforestation projects lead to a positive change such as the formation of clouds. These clouds then reflect the sunlight, lowering temperatures.

Planting trees in tropical climates with wet seasons has another advantage. In such a setting, trees grow more quickly (fixing more carbon) because they can grow year-round. Trees in tropical climates have, on average, larger, brighter, and more abundant leaves than non-tropical climates. A study of the girth of 70,000 trees across Africa has shown that tropical forests fix more carbon dioxide pollution than previously realized. The research suggested almost one fifth of fossil fuel emissions are absorbed by forests across Africa, Amazonia and Asia. Simon Lewis stated, "Tropical forest trees are absorbing about 18% of the carbon dioxide added to the atmosphere each year from burning fossil fuels, substantially buffering the rate of change."

As of 2008 1.3 billion hectares of tropical regions were deforested every year. Reducing this would reduce the amount of planting needed to achieve a given degree of mitigation.

Incentives

Some incentives for reforestation can be as simple as a financial compensation. Streck and Scholz explain how a group of scientists from various institutions have developed a compensated reduction of deforestation approach which would reward developing countries that disrupt any further act of deforestation. Countries that participate and take the option to reduce their emissions from deforestation during a committed period of time would receive financial compensation for the carbon dioxide emissions that they avoided. To raise the payments, the host country would issue government bonds or negotiate some kind of loan with a financial institution that would want to take part in the compensation promised to the other country. The funds received by the country could be invested to help find alternatives to the extensive cutdown of forests. This whole process of cutting emissions would be voluntary, but once the country has agreed

to lower their emissions they would be obligated to reduce their emissions. However, if a country was not able to meet their obligation, their target would get added to their next commitment period. The authors of these proposals see this as a solely government-to-government agreement; private entities would not participate in the compensation trades.

ECOFORESTRY

Ecoforestry has been defined as selection forestry or restoration forestry. The main idea of Ecoforestry is to maintain or restore the forest to standards where the forest may still be harvested for products on a sustainable basis. Ecoforestry is forestry that emphasizes holistic practices which strive to protect and restore ecosystems rather than maximize economic productivity. Sustainability of the forest also comes with uncertainties. There are other factors that may affect the forest furthermore than that of the harvesting. There are internal conditions such as effects of soil compaction, tree damage, disease, fire, and blow down that also directly affect the ecosystem. These factors have to be taken into account when determining the sustainability of a forest. If these factors are added to the harvesting and production that comes out of the forest, then the forest will become less likely to survive, and will then become less sustainable.

Since the forest is considered an ecosystem, it is dependent on all of the living and non-living factors within itself. This is a major part of why the forest needs to be sustainable before it is harvested. For example, a tree, by way of photosynthesis, converts sunlight to sugars for respiration to keep the tree alive. The remains of the converted sugars is left in roots for consumption by the organisms surrounding the tree in the habitat. This shows the productivity of an ecosystem with its inhabitants. Productivity within the ecosystem cannot come to fruition unless the forest is sustainable enough to be harvested. If most individual organisms of the ecosystem vanish, the ecosystem itself is at risk. Once that happens, there is no longer any forest to harvest from. The overall productivity of a system can be found in an equation where the Net Primary Production, or NPP, is equal to the Gross Primary Production, or GPP, minus the Respiration, or R. The formula is the $NPP = GPP - R$. The NPP is the overall efficiency of the plants in the ecosystem. Through having a constant efficiency in NPP, the ecosystem is then more sustainable. The GPP refers to the rate of energy stored by photosynthesis in plants. The R refers to the maintenance and reproduction of plants from the energy expended.

Ecoforestry has many principles within the existence of itself. It covers sustainable development and the fair harvesting of the organisms living within the forest ecosystem. There have been many proposals of principles outlined for ecoforestry. They are covered over books, articles, and environmental agencies. All of the principles relate to the idea that in ecoforestry, less should be harvested, and diversity must be managed.

Through harvesting less, there is enough biomass left in the forest, so that the forest may stay healthy and still stay maintained. It will grow at a sustainable level annually, and thus it will be able to still be harvested the following year. Through management of the diversity, species may cohabitate in an ecosystem where the forest may feed off of other species in its growth and production.

Principles of Ecoforestry

The principles of ecoforestry are:

- One must consider retention: It must be the first consideration in any planned removal of trees from a stand. By placing retention at the beginning of a planned removal, one makes sure that they know what must be left to ensure the protection of such things as rare species, sites of native cultural significance, riparian zones.
- Riparian zones should not be touched: It is a sensitive area, thus tree removal should not occur there. This protects the water quality. Water quality is protected by not altering the drainage pattern of the zones. No tree removal should take place in the most sensitive areas.
- The composition and structures should be upheld so that forests may fully function. This may be subject to large old trees, snags, and large fallen trees. These pieces of the ecosystem are upheld through letting them grow and die out into a timber extraction area.
- When removing trees, the lowest impact should be used: This means not compacting the soil in the forest and building small or no roads if at all possible. This takes away the disturbance that would be done to the foundation of the forest's grounds.
- Plan in terms of the needs of the larger watershed: The watershed zone plan designates areas where tree removal is not permitted. It also designates the areas where removal is possible, and the different types of removal available.
- Prohibit clear cutting: Clear cutting is not ecologically correct. There are methods for cutting trees to maintain structure, and to allow the trees to cohabitate in a healthy ecosystem.
- Select trees as candidates for removal by considering how abundant and redundant their structures and functions are to the rest of the forest as a whole, leaving potential wildlife trees (to become snags and coarse woody debris).
- The forest should be allowed to regenerate trees through the seeds from the trees in the logged areas. This allows tree planting to be taken out of the ecoforestry principles, allowing a natural regeneration.

- Ecological succession should be kept at all times. This will protect biological diversity. This even means eliminating the process of brush control.
- No slash burning: Fire may be used as a tool in landscapes that have a history of naturally occurring fires. Although this may be true, fire should still be used with caution.
- No pesticide use: The forest needs disease, insects, and shrub/herb vegetation. They are essential parts of a fully functioning forest, even though they involve the forest's decay in some cases. They are natural parts that allow the ecosystem to operate as a whole.
- Maintain and restore topsoil quality: This can be done through leaving sufficiently coarse and small debris on the forest grounds.
- Maintain beauty and other natural aesthetic qualities in the visual, sound, and odor landscapes. This includes not taking away the wildlife, the plants, or structural layout of the ecosystem.
- Always look at the forest as a whole: Each part of the forest contributes to its overall needs and health. That is how the forest would have survived without human interference.
- Rely more on people and markets: Use accounting and budgeting as a solution to rely less on the destruction or harvesting from the forests.
- Don't do wrong: If it feels wrong, then it is probably wrong. Don't allow ignorance to persevere. Recognize that the ecosystem needs tending to. If the forest is not preserved, then it cannot be harvested forever.

ANALOG FORESTRY

Analog forestry is an approach to ecosystem restoration that considers the process of forest formation and the functioning of forest services to be critical in establishing a sustainable ecosystem characterised by a high biodiversity to biomass ratio. Design is produced through a synthesis of traditional and scientific knowledge. It seeks to optimise the productive potential of the design rather than maximise the production of one crop and to maximise ecosystem services by increasing the volumetric mass of the photosynthetic component.

Analogue forestry draws design input not only from traditional models but also from the natural forest successional dynamics. When an ecosystem is designed to be analogous to the indigenous climax state, the efficiency and dynamics of the natural processes can be replicated. These quasi-natural forests are designed to mimic the structural and functional aspects of indigenous forests and are referred to as analog forests. In addition to their ecological characteristics, analog forests are also designed to provide economic benefits. However, it is not until all the ecological requirements of the location are satisfied that economic values of species are considered. Therefore, an analog forest may comprise natural and exotic species in any proportion, the contribution to structure and function being the overriding factor that determines its use.

The theoretical underpinnings began in 1978 in San Diego and Guatemala. It was first implemented in Sri Lanka around 1981 by Ranil Senanayake as an alternative to monocultures of *Pinus* and *Eucalyptus* and has spread to India, Vietnam, Philippines, Australia, Peru, Ecuador, Colombia, Brazil, Costa Rica, Dominican Republic, Honduras, Mexico, Canada, Kenya and Zimbabwe at present.

The International Analog Forestry Network (IAFN) is currently hosted in Costa Rica.

Analog forestry is a system of silviculture that seeks to design, plan and manage ecosystems dominated by woody perennials. It has been primarily employed in tropical or subtropical areas, but can be used in temperate areas too. The design seeks to mimic the architectural structure and ecological function of the preexisting climax vegetation of the area, and can be designed to provide economic, social and environmental benefits.

Analog forestry always considers the inclusion of non-crop native species to be considered in the design and thus addresses the inclusion of native biodiversity in the design. As analog forestry also requires the inclusion of long-lived species of trees in the design it has the capacity to sequester carbon for a longer time than plantation forestry. Analogue forestry has the potential to produce very high values of photosynthetic biomass as its design calls for the inclusion of all the growth forms that occupy the three-dimensional space of the mature indigenous forest. By including many species of crops in one area analog forestry helps spread the risk of market failure on a single crop.

MYCOFORESTRY

Mycoforestry is an ecological forest management system implemented to enhance forest ecosystems and plant communities through the introduction of mycorrhizal and saprotrophic fungi. Mycoforestry is considered a type of permaculture and can be implemented as a beneficial component of an agroforestry system. Mycoforestry can enhance the yields of tree crops and produce edible mushrooms, an economically valuable product. By integrating plant-fungal associations into a forestry management system, native forests can be preserved, wood waste can be recycled back into the ecosystem, planted restoration sites are enhanced, and the sustainability of forest ecosystems are improved. Mycoforestry is an alternative to the practice of clearcutting, which removes dead wood from forests, thereby diminishing nutrient availability and reducing soil depth.

Selection of Fungal Species

According to Paul Stamets, the first principle for the creation of a mycoforestry system is to utilize native fungal species. Implementing a mycoforestry system provides the potential of improving restoration efforts and the possibility of economic gain through mushroom cropping and harvesting. However to utilize native fungal flora, first the relationships between present fungal species and growth substrate, and habitat need to be studied.

A simple way to introduce a mycoforestry system and enhance out-plantings for crops and forest restoration sites is to "use mycorrhizal spore inoculum when replanting forest lands" For this process it is best to match native trees with native mycorrhizal fungi. This method keeps and will promote the functioning of the native ecosystem, and native biodiversity.

It is assumed in a functioning forest ecosystem an underground mycelial network persists even if no fruiting bodies are visible. A period of disappearance of mushrooms from an area should not cause alarm. In order to trigger the formation of fruiting bodies, many fungal species require specific environmental conditions. Most species of fungi do not fruit year round.

Mycoforestry is an emergent scientific field and practice. Until broadly standardized protocols are created and perfected, the collection of both current and historical ecological site conditions will improve the success of the project. Therefore, a survey of fungal relations at the site under both prime and poor conditions is beneficial to implementation of a mycoforestry system.

Saprotrophic Fungi

Edible oyster mushrooms (*Pleurotus* sp.) fruiting from a stump.

The second principle is to promote saprotrophic fungi in the environment. Saprophytic fungi are crucial to mycoforestry systems because these are the primary composers breaking down wood and returning nutrients to the soil for use by the rest of the forest ecosystem. This can be accomplished through inoculation of wood debris at site. Spored oils can be used in chainsaws when problematic or invasive hardwood requires felling. This method is a simple means to inoculate a tree. Additionally plug spawn can be implemented and injected into wood mass again prompting colonization by the selected fungus. Eventually repeated colonization efforts should not be necessary as many fungal life forms are strong and will spread and sustain in the soil on their own.

In management of the mycoforestry system it is important that dead wood be in contact with the ground. This allows fungus to reach up from the soil and decompose fallen wood releasing nutrients at a much quicker rate then if the wood is left standing. Additionally it is important to leave dead wood on site for decomposition back into the soil. This philosophy is similarly based to the fact that clear cutting of a forest reduces soil nutrients and thickness.

Beneficial Fungal Interactions

Armillaria, a parasitic fungus.

The third principal is to implement species known to benefit plant species. These are commonly mycorrhizal fungus that form long term associations with plants, often extending inside of plants roots acting as an additional root system providing for better absorption of nutrients and water.

Utilizing mushroom species that attract insects could be a useful source of fish food. This practice makes the mycoforestry a larger system. Unlike most agriculture systems it helps the environment in a number of ways. It ties all biological aspects of the environment together, creating sustainable living and food production as well as sustainable fisheries similar to the ancient Hawaiian Ahupua'a, which utilized sustainable all portions of the land for environmental and food security.

Additionally fungal species can be implemented that compete with disease causing agents like Armillaria root rots to provide long term protection of the forestry system.

Additionally, the implementation of an agroforestry system performs mycoremediation and mycofiltration activities cleaning up toxins and restoring the environment.

WILDLIFE FORESTRY

Wildlife forestry is management of forest resources, within sites and across landscapes, to provide sustainable, desirable habitat conditions for all forest-dependent (silvicolous) fauna while concurrently yielding economically viable, quality timber products. In practice, however, management decisions associated with wildlife forestry often reflect a desire to provide suitable habitat for rare species, species with declining populations, and exploitable (i.e., game) species. Collectively, these species are deemed priority species and they are assumed to benefit from habitat conditions that result from prescribed silvicultural management actions.

Early wildlife conservation efforts largely focused on controlling indiscriminate slaughter of wildlife by restricting the season or sex of harvested species. Subsequent conservation efforts targeted increased protection of populations through creation of sanctuaries or reserves, such as national parks and wildlife refuges, within which harvest of wildlife was prohibited or greatly restricted. Typically parks and wildlife reserves were located in areas with abundant wildlife populations, and therefore, little emphasis was placed on active management to improve or maintain suitable habitat.

In his description of a "land ethic" in which people are members of the natural community, Leopold indicated the appropriateness of habitat alteration and management to aid the continued existence of wildlife species. His philosophical principles were succinctly stated as "A thing is right when it tends to preserve the integrity, stability, and beauty of the biotic community. It is wrong when it tends otherwise."

Anthropogenic alteration of forest habitat for the benefit wildlife species has long been practiced, particularly with regard to intentional use of fire. Even so, active manipulation of forest structure to purposefully enhance wildlife habitat has been slow to evolve. Most early efforts regarding habitat alteration focused on providing suitable escape cover, foraging habitat, or distinct habitat features (e.g., nest sites) for game species. Within forested landscapes, improving wildlife habitat often meant increasing local heterogeneity and providing more forest edge habitat. However, increased knowledge of the possible negative effects associated with forest edges, such as greater predation and nest parasitism rates, has called into question the practice of providing increased heterogeneity within forested landscapes. At the same time, a desire to provide habitat for specific species that may require specialized habitat conditions, as is the case for some endangered and threatened species as well as other charismatic megafauna with widespread public support, has prompted development of habitat management prescriptions designed to improve forest habitat conditions for specific species.

Partly in response to what has been perceived as management for single-species, but also to advocate a desire to prevent future population declines of common species, recent emphases in wildlife conservation have been placed on the development of comprehensive, eco-regional, conservation plans whose scope encompasses multiple species or the entirety of a species-group. This approach gained momentum with the North American Waterfowl Management Plan which identified conservation actions intended to return waterfowl populations to their former abundance and distribution. Subsequently, this basic concept has been expanded to other species groups such as songbirds, reptiles and amphibians, and shorebirds, as well as other species groups. Although these conservation plans differ markedly with regard to their recommendations for habitat management, many of these conservation plans advocate use of prophylactic management prescriptions to enhance or maintain suitable habitat conditions for priority wildlife species. Indeed, most conservation plans recognize the need for alteration of habitat and landscapes via management actions so as to attain the desired distribution, abundance, and viability of priority wildlife species.

Before managers can provide appropriate forest habitat for priority wildlife, the landscape and site characteristics that contribute to viable populations of these species must be identified. Preferably, those forest habitat characteristics identified can be quantified via standard measurement protocols. If possible, the desired or acceptable range of values should be determined, and either quantitatively or qualitatively stated, for each of these characteristics. In addition, a threshold value for each characteristic should be determine, that if exceeded would justify prescription of management actions to return site characteristics within the desired range.

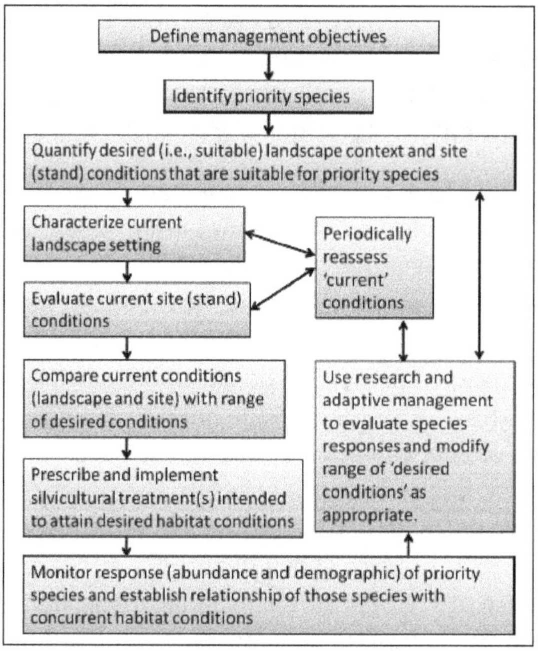

Characterization of wildlife forestry silvicultural methodology.

Because wildlife forestry silviculture does not target optimal production of forest products, many forest management practitioners (i.e., foresters) have been reluctant to adopt these alternative silvicultural practices. Indeed, these concerns may be justified, as desired stand conditions may include less then fully stocked stand densities, retention of some economically mature trees, maintaining species diversity which includes less merchantable tree species, and preservation of less vigorous (decadent) trees. However, at present we lack credible, long-term economic data to suggest that development and maintenance of quality forest habitat for priority wildlife species will result in a sacrifice in timber production or quality. Regardless of these overarching financial concerns, silvicultural management actions prescribed to enhance desirable forest habitat conditions for priority wildlife species are expected to be commercially viable. Maintaining the long-term commercial viability of prescribed silviculture is import, because without sufficient financial returns, needed management actions will not be undertaken.

Forest Management

Silviculture is the manipulation of the establishment, growth, composition, and quality of forest vegetation to achieve management objectives. Historically a primary objective on managed forest lands has been to maximize financial returns from production of forest products such as timber and wood pulp. Production-oriented silviculture has focused on reducing the length of production cycles between stand regeneration and harvest while concurrently maximizing wood product yields. Reduction of production cycles (i.e., rotation length) tends to lessen the time regenerating forests are in early seral habitats, and consequently the size and age of shrubs and there their likelihood of persistence within subsequent seral stages is reduced. Often production-forestry objectives favored a limited number of tree species due to their high economic value or growth potential. Concomitantly, to increase growth and yield of desired trees, other species with lesser economic value, as well as congeneric competitors, were culled. These actions result in increased spatial homogeneity within managed stands. The apogee of this trend towards increased spatial regularity is expressed within planted plantation forests. Climate, soils, species, and other factors affect the optimal length of production cycles within productionoriented forests. Thus, rotation periods within plantations vary markedly from rotations of up to 150 years to as few as 10 years. Notably, recent developments and increased advocacy of short-rotation woody coppice systems for use in biomass production have further compressed the length of production cycles to<4 years.

Even in non-plantation forest management systems, historical forest management practices, especially those focused on timber or pulpwood production, have often resulted in relatively homogeneous mature forest conditions. This homogeneity has been exacerbated by environmental changes that limit natural disturbances within forest stands (e.g., fire suppression). When these homogenous forests are managed using even-aged silviculture, during which all or most trees are harvested and the entire stand regenerated, successional 'boom and bust' conditions are dispersed throughout landscapes. The forest conditions resulting from even-aged management (e.g., clear cuts) are temporarily favorable for some species, specifically benefiting species that exploit early-successional forest habitat. Even so, these early-successional habitats are transitory and are followed by an extended period of stand development during which the homogeneous, closed-canopy forest structure with sparse understory vegetation offers attraction (forage or cover) primarily for common (non-priority) forest wildlife species. Indeed, the abundance of homogeneous, closed canopy forests with sparse understory cover that have resulted from historical silvicultural practices may have contributed to why species for which these habitat conditions are suitable are considered common and are not priority species in need of management actions to increase their abundance. In addition, the altered (often depauperate) species compositions that have resulted from selective harvest and culling during past management practices may amplify the potential for catastrophic damage from outbreaks of insects or diseases or limit the seasonality of mast crops that are exploited by wildlife.

In contrast to the relatively short periodicity of production-oriented silviculture and the periodic boom-bust cycles associated with even-aged silviculture, natural successional processes within forests may require hundreds, or potentially a thousand, years to achieve complete replacement via small-scale disturbance or as the duration between catastrophic stand-replacement events. Owing to differences in their successional development, structural and ecological characteristics within production-oriented managed forests and in forests managed using even-aged silviculture are manifestly different from those same characteristics within natural forests. Ecological differences between managed and natural forest stands are exacerbated over time, as natural stands approach climax or 'old-growth' conditions while intensively managed forests are repeatedly regenerated.

To mitigate the detrimental effects of forest management on priority wildlife species, recommendations have been made to increase the length of production cycles, thereby allowing increased time for development of heterogeneous forest structure. Unfortunately, small increases in rotation lengths provide little opportunity for increased complexity of forest structure within these stands – and periodic, complete harvest and stand regeneration thwart retention of forest structure over time. Similarly, forest managers have argued that temporal and spatial dispersion of clear-cuts throughout landscapes will provide continual availability of wildlife habitat – albeit at different locations. Although early-successional habitat may be continuously afforded via such dispersion, few of the other benefits conferred by increased forest structural heterogeneity, such as emergent or senescent legacy trees, are realized.

Increased recognition of the structural and ecological deficiencies within production-oriented forests has spurred interest in and development of alternative silviculture. Alternative silviculture has been especially valuable for stands where management objectives are not solely financial but where stewardship objectives include preservation of regional biodiversity, enhancement of wildlife habitat, maintaining landscape aesthetics, or providing increased recreational opportunities.

Alternative silvicultural methods developed for use in many forest types have been based on identification and quantification of natural disturbance regimes, with subsequent implementation of silvicultural methods intended to emulate these disturbances. For many forest types, emulation of natural disturbance regimes includes small-scale disturbances implemented via single-tree and patch-cut harvests. If successfully implemented, emulation of natural disturbance regimes should promote natural stand development and succession. Unfortunately, natural stand development and concurrent successional changes may conflict with landowner objectives if succession results in changed species composition or reduced financial returns. Moreover, strict adherence to silvicultural regimes that mimic natural disturbance may not be necessary to maintain biodiversity or to provide desired wildlife habitat.

Another common alternative silviculture has been inspired by the desire to achieve forest structure reminiscent of old-growth forest. Although the structure of old-growth forest varies among forest types, common attributes generally include increased vertical heterogeneity via a multi-layered canopy, increased horizontal heterogeneity associated with canopy gaps and different regeneration cohorts, the presence of large trees that are often in older-age classes, abundant snags (standing dead trees) or senescent trees, and large diameter downed woody debris. Myriad ecological benefits, including increased biodiversity, have been ascribed to old-growth forests. As such, forest reserve lands have been designated (e.g., wilderness areas and parks) which afford unfettered, natural development of these characteristics. However, on lands with timber production as an objective, even if not the sole objective, implementation of production cycles that span hundreds of years is not likely to occur. In addition, merely increasing the length of production cycles may provide little enhancement of biodiversity if the forest structure favored by production-oriented silviculture is maintained. Furthermore, where forest reserves are small or located within inhospitable landscapes, full ecological benefits of old-growth forests may not be attainable or sustainable.

Fortunately, it is likely that many of the benefits conferred by old-growth forests, or at least specific species associated with old-growth forests, are positively related to the heterogeneous structure (vertical and horizontal) and other structural attributes found within these forests, not necessarily with the prolonged existence of these forests. As such, alternative silvicultural methods have been developed which retain or enhance structural heterogeneity within managed forests, thereby promoting structural attributes of old-growth forests and encouraging greater biodiversity. Most often, increased heterogeneity and other desired attributes have been achieved via retention of living or dead trees (or other forest elements). These retained structures have been referred to as legacy elements.

Based on the concept of retained legacy elements, various alternative silvicultural methods have been advocated to enhance structural complexity within managed forests. Some of these methods have being designated as green-tree retention harvests, variable-retention harvests, dispersed retention harvests, variable density thinning, and active intentional management. The legacy elements retained under these different silvicultural regimes may differ markedly. Retained elements may differ in quantity (e.g., % canopy cover or % basal area retained;) but they may also differ in their composition, as different canopy species are preferentially harvested or specific regenerating species favored for retention. In addition, the quality of retained elements may differ via selection of retained trees with differing perceived fates – from robust, canopy-emergent trees through decadent, dead or dying trees. Moreover, retained elements may differ in their dispersion (e.g., evenly dispersed or aggregated;) and the spatial juxtaposition of retention.

Wildlife forestry silviculture exploits the versatility of legacy retention by combining specific retention and removal of canopy to achieved habitat conditions that are deemed suitable for priority wildlife species. The resultant forest stand may exhibit a mixture of tree species, size classes, and decadence within a heterogeneous distribution of aggregated retention and canopy gaps.

Landscape Considerations

Although specific silvicultural methods have most often been developed for and are implemented within forest management units (a.k.a. forest stands), a similar ecologically based approach has evolved with regard to maintaining or restoring landscapes surrounding managed forest stands. Intact forested landscapes are conducive to maintaining floristic and wildlife diversity, but also provide tangible social and economic benefits.

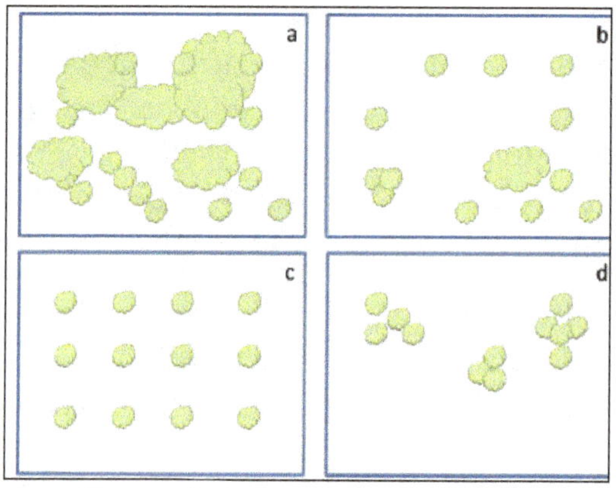

Potential distribution of legacy elements after application of alternative silviculture, differing in quantity of retained canopy cover (a, b) or dispersion of retained elements (c, d).

Loss of diversity within landscapes is often a reflection of the increased homogeneity within forest stands. Although forest structure within a landscape ranges from regenerating stands to mature merchantable stands, these stands may be monotypic in species composition and physical structure. In addition, ensuring appropriate size of habitat patches and maintaining connectivity among these patches is critical to sustaining viable wildlife populations within landscapes. Although the appropriate patch size is dependent upon the species under consideration, increased fragmentation tends to disrupt dispersion of fauna and flora and when habitat area is reduced without appropriate connectivity, extirpation of some species can result. Finally, maintaining landscapes capable of sustaining viable wildlife populations requires that evaluation, planning, and management decisions reflect the temporal processes inherent in the ecosystem. Four landscape conditions have been recognized as contributing to ecosystems that support sustainable wildlife populations: 1) diversity within the landscape, 2) connectivity among landscape elements, 3) appropriate habitat patch sizes, and 4) sufficient time to achieve ecological functions. The impact of the degradation of forest landscapes was succinctly summarized by Franklin as: "In general, we have tended to forget that what is good for wood production is not necessarily good for other organisms or processes in a forest ecosystem. Fully stocked young forests,. are the most simplified stage of forest development in terms of structure and function, and the most impoverished in terms of biological diversity. Simplification-genetic, structural, landscape and temporal—reduces ecosystem resilience. Thus the key to retaining resilience must be in maintaining ecological complexity or diversity".

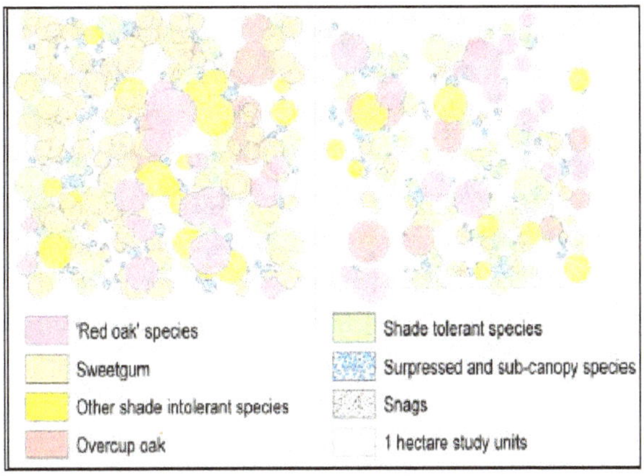

Canopy cover accounted for by individual tree crowns within bottomland hardwood forests on an untreated control stand (left) and on a stand subjected to wildlife forestry variable-retention silvicultural treatment (right), 6 years after treatment.

As when considering disturbance within the context of forest stand development, landscape conditions have often been influenced by large-scale disturbance events, particularly fires. Forest landscapes that result from management based on historical fire regimes, when compared with landscapes with extensive reserves and traditional silvicultural management, had increased area of late-successional habitat, more overstory structure in young forests, and larger forest patches. Similarly, when compared with human altered landscapes, natural landscapes were composed of larger habitat patches which formed the landscape matrix, had greater connectivity among patches, the range of variation in patch size and complexity was greater, as was the likelihood of encountering 'rare' habitat types.

Despite some understanding of desirable characteristics of forested landscapes, the appropriate landscape area remains nebulous. In part this is because an appropriate landscape area is dependent upon the species being considered, landscape characteristics, desired response, and the management context. As an example, the landscape appropriate for a grizzly bear is vastly different from the appropriate landscape for a shrew.

Even with knowledge of the appropriate landscape area and characteristics, manipulation of landscape conditions is usually beyond the purview of individual landowners. This is true even for large industrial or governmental landowners. Nevertheless, landscape conditions influence the ecological benefits conferred by forest stands, including the distribution, abundance, and viability of wildlife species within forest stands. In boreal forests, stand structure accounted for 54% of variability in bird and small mammal communities but landscape characteristics accounted for 29%. Similarly, bird species abundance within coniferous dominated landscapes was positively related to the proportion of broadleaf forest, with landscape thresholds for individual species ranging from 1% to 25% canopy cover. A strong negative relationship was noted between landscape broadleaf forest and avian population trends, such that the species most associated with broadleaf forests had greatest population declines. Therefore, maintaining or restoring broadleaf vegetation is important within coniferous dominated landscapes.

Because of landscape influences, providing desired forest conditions within forest stands may be insufficient for some priority species if these stands are located within inhospitable landscapes. Moreover, for long-term sustainable populations, most sites, even those with suitable habitat conditions, must have spatial and temporal linkages that facilitate dispersal, colonization, and interchange within metapopulations. Towards that end, desired landscape conditions for priority wildlife species are usually composed of large forest patches as the preponderance of the landscape matrix, have a wide range of variation in patch sizes and complexities, and provide extensive connectivity among forest patches.

Why Wildlife Forestry?

The application of wildlife forestry silviculture is in many respects the logical extension of the various alternative silviculture methods whose objectives are to maintain or increase forest biodiversity. Although forest management targeting increased biodiversity is a laudable objective in and of itself, managers of public conservation lands (e.g., National Wildlife Refuges and State Wildlife Management Areas) or private lands under conservation easements may have legal, or self-appointed, mandates to manage and conserve priority wildlife species rather than the more general goal of improving biodiversity. On other private lands, the economic and recreational values associated with silvicolous wildlife, including fee hunting, hunt-club memberships, wildlife viewing and other eco-tourism activities, have income potential that may equal or exceed that of extractive timber harvest. Hence, ensuring sustainable populations of wildlife may be essential for maintaining financial revenues from these forest lands.

In addition, land ownership patterns of private forest lands are changing. Changing ownership demographics and associated alteration of management objectives of these forest land owners have resulted in smaller landownership holdings. For owners of small landholdings, maintaining continuous forest cover and the aesthetic considerations associated with forests may be of paramount importance.

Effects of Wildlife Forestry and other Silviculture on Stand Structure

Although not specially referring to wildlife forestry, Carey succinctly encapsulated the holistic approach of wildlife forestry as - “Management of forest development and landscape dynamics is more likely to be successful in maintaining ecosystem and landscape function than just providing select structural elements in stands and select structural stages in landscapes, as is often suggested for conservation."

Treatment	Increasing forest diversity, structure, and complexity		
Regeneration harvest	Clearcut	Legacy retention; dispersed or aggregated	Wildlife forestry; variable retention with canopy gaps
Restoration planting	monoculture	Monoculture with supplemental natural reproduction	Mixture of species; hard and soft mast; shade tolerant and intolreant
Stand modification (TSI)	Herbicide removal of understory and midstory	Partial removal of competitors	Heterogenous retention and removal of understory competitors
Thinning	Systematic single species retention	Multi-species retntion	Heterogeneous retention and thinning; underrepresented species preferably retaind

CLIMATE CONTROL AND SUSTAINABLE FOREST MANAGEMENT

By accepting that environmental conditions are *dynamic and therefore uncertain*, we hypothesize that the adaptability (adaptive capacity) of the forest and the forest enterprise are of more concern, and will lead to forest structures other than those obtained from deterministic approaches that aim for adaptedness.

Historically, in forest management and in forest science, there is much less experience with structures that ensure high adaptability while pursuing the forest owner's goals than there is experience with measures for adaptedness to constant environmental conditions.

Some concepts which were established in central Europe at the end of the nineteenth century can be seen as an effort towards creating more flexible and adaptable forests. These concepts began with Gayer's call for mixed stands. Later Möller developed a truly holistic concept. He considered the forest to be an organism and suggested that the continuity in woodland conditions over time should be a central part of forest management practice. Still, there are not many examples of true management approaches to high adaptability. The idea of acting explicitly to promote sustainability in order to "open up service opportunities in the future" in forests is becoming established very slowly.

However, adaptability also means an economically flexible forest. An example of this approach for achieving flexibility in the economical sense is to spread the risk of investments by keeping a high diversity of species in order to be well positioned in any market situation. This approach follows the portfolio theory.

Millar have pointed out that, essential to managing forests in the face of uncertainty, is the imperative to learn as you go. These authors encourage flexible approaches that promote reversible and incremental steps, and favor ongoing learning and the capacity to change direction as situations change. These approaches closely resemble what economists have termed ecological rationality, where initially constructivist designs are modified in the light of test results, are tested again, and are retested and so on to effect an evolutionary adaptation. With this approach, the forest is seen as a laboratory. This strategy was best described by Walters and Holling for the management of renewable resources and has gained widespread acceptance so far.

A maximum flexible forest may be described, theoretically, by its ability to adapt to any change in the ecological or economic environments. In this view, ecological functions, as the basis for the functioning of systems in a changing environment, are set prior to economic and social ones. However, the extent to which the economic and social functions might follow depends on many factors, among which the aims and expectations of the different owners are surely the most relevant. We may think of admixtures of economically interesting tree species being deliberately introduced to the site to increase economic flexibility in one case, and, in another case, solely relying on a site-specific native tree species assemblage to provide habitat for many organisms. Still, this is a radical change from the deterministic approach which has characterized forestry until recently and which is based on the wake theory; that theory states that if forests are efficiently managed for wood production, then all the other forest functions will follow. The specific services this maximum flexible forest will provide are not those attained from the short-term management aim, but, on the contrary, they are byproducts of complex forest ecosystem functions.

What Ciancio have termed systemic silviculture may provide a description of such a forest. With systemic silviculture, the cultivation unit is the stand. Silvicultural and utilization interventions are cautious, continuous, and capillary in relation to the needs of the various stands. The forest is inhomogeneous. There is no predefined rotation age. Regeneration is natural and continuous. Monitoring is an essential element for adapting cultivation and management to the responses of the system. Managers aim towards conserving and increasing complexity. Wood production (and other services) is a byproduct of cultivation carried out in favor of the ecosystem's functionality.

Finally, management based on the systemic silvicultural approach implies decentralized control. To this end, systemic silviculture shares ideas of adaptive forest management) in that feedback loops are an essential part of both concepts. Furthermore both concepts aim to make forests and forest enterprises adaptive and responsive to changing conditions.

The concept that a "minimum" growing stock, which should always be present on a management unit, as suggested by Ciancio , could answer both the question of keeping forest ecosystems above the "critical" zone, where the ecosystem may transfer to another stability domain , and the question of keeping more options open for the future provision of ecosystem services (i.e., increase flexibility).

Management Response to Climate Change under the Adaptable Forest Approach

The indeterministic approach of coping with climate change would involve adding flexibility (improved suitability of a forest ecosystem to be adapted to demands in the unknown future), resilience, and capacity for self regulation to the forest. These issues have long been of general concern in ecology.

Site characteristics and vegetation zones have to be taken into account when adaptability is the aim; this is also true when adaptedness is the aim. This means that it is very difficult to provide general ideas of how to structure and treat a forest ecosystem for the intended management goals.

However, identifying the most important ecosystem features for allowing for high resilience and self-regulating capacity is also an aim of decision-oriented sciences such as silviculture. This became obvious when existing management systems fell under environmental pressure, as was the case in the 1980s in central Europe due to air pollution. Ulrichsummarized ecosystem features relevant for forest resilience in a hierarchical system with regard to emissions and climate change. Many of the ideas established during that time, which were founded in the ecosystem sciences, have led to operative measures for forest transformation in central Europe.

However the aims of the indeterministic concept go beyond just imitating nature. Although there are few examples of manager's approaches for high adaptability with regard to climate change, we believe that the following paths should be given some thought.

- Ecological knowledge suggests specific structural elements, e.g., aggregates of conspecific trees and structural features, such as gap frequency distributions, gain importance as management treatments are defined by the monitoring of changes in such indicators. The regenerative capacity of forests may be taken as an example: Indeterministic management would increase tree species diversity, perhaps even beyond the natural range, and ensure natural regeneration of those species. This may foster self regulation in the case of disturbance events. Self-regulation ability could be promoted by small proportions of different species being arranged in aggregates and positioned at distances that correspond to the pollen-and-seed-dispersal abilities of the species. This pattern may allow for an effective natural regeneration of each species within a given area and, simultaneously, decrease the need for intense forest management. For pioneer species such as birch, the proportions could be very small, e.g., less than 2%, and the distances between the aggregates could be fairly large, e.g., some 100 m. This would still allow for regeneration across the area if the dominant stand were eliminated e.g., by a storm. In contrast, shade-tolerant species with low dispersal ability may be maintained in the stand understory, to become reproducing individuals after release.
- As part of the high-adaptability strategy, the fitness, vitality, and regeneration

capacity of tree species populations become important issues. Therefore, it is important to consider that pollination and fructification depend on crown surface area as well as on general growth parameters. Thus, release treatments are a useful means for guaranteeing vitality and fitness of the highest possible number of tree species. This applies all the more because intensive mixtures of different tree species at the stand level (shade-intolerant *versus* shade-tolerant) lead to high competitive pressure.

- To maintain the different species in the forest permanently, a variety of regeneration cuts could be applied to produce small to large gaps, e.g., by single selection, group selection, and edge cuts. In particular those beech ecosystems of central Europe in which disturbance is rare and species diversity is minimal may be shifted to systems which host more (tree) species by emulating a more disturbance-intensive management system. This follows Connell's very general theory on the promotion of species richness through medium disturbances. This aspect makes a difference to traditional "close-to-nature" forestry practiced in central Europe where the preference for small disturbances—i.e., mimicking the natural gap size frequency distribution of untouched beech forests—is often combined with target diameter cuttings.
- Aside from the general idea of promoting species diversity, the idea of buffering environmental stress by tightly coupling the primary and secondary production in space and time is advocated in soil science. Ulrich stated that ecosystems with " soils with a good buffering capacity with respect to changes in their nutrient, acid-base, water, and oxygen status can withstand a greater variance of the climate ." This was stated explicitly for events such as "dry periods, acidification, and nutrient losses". Combined with the establishment and maintenance of a diverse herbaceous layer, the choice of tree species is, again, crucial. However, species choice and mixture may be guided by tree species litter decomposition and rooting depth traits. Nitrogen-fixing species or species well known for their ability to selectively sequester cations come to mind.

The adaptable forest approach will not suit all forest owners because it may mean accepting trade-offs between the level of provision of the different ecosystems services from the forest in the short run. As an example, for a private forest owner who aims at maximizing returns from wood production, an increase in adaptability could mean reducing, at least in the short run, the level of production. On the other hand, on public properties where the aim is to maintain life-supporting ecosystem services (soil conservation, watershed protection, carbon storage, etc.) such an approach will increase the probability that the forest will continue to provide these functions and services even if its structure and composition change following adaptation.

What Accompanies this Approach?

The main difficulties with this type of approach arise from three factors: (1) The positivist view, which is still firmly ingrained in forest science; (2) The fact that most managed forest ecosystems have been managed for a very long period of time and show structures and functions that have generally been simplified by managers; and (3) Institutional settings with pre-existing goals and rules. The first point makes it difficult for many foresters to accept that it could be possible to manage a forest without trying to match reality to expectations. This is not to say that there is no target stand structure. However, there is no longer a single stand target structure. On a larger spatial scale we have already identified important structural features. At least the idea of having multiple structures is concrete.

The second point can be very important when dealing with very simplified forest structures, such as even-aged, monospecific stands which have a very narrow range of adaptability, thus making it more difficult and time consuming to increase stand diversity and expected adaptability. Moreover there is the awareness that the rate of climate change might override any adaptive capacity of present-day forest ecosystems. Furthermore, increasing adaptability could mean reducing the rate or level of products and services coming from the forest.

Concerning the third point, forestry has a long history of rules and regulations aimed at avoiding the reduction in a forest's capacity to fulfill different services through excessive or irrational use. To adopt an integrally adaptive approach in forest management might entail counteracting or not entirely fulfilling the requirements set by specific regulations. An example could be the unintentional shift in species composition and habitat structure in areas where maintaining existing biodiversity is an institutional goal. Or another example is the longer time that may be needed for natural regeneration to spread where, instead, existing rules might stipulate that new trees fill in utilized areas within a defined period of time.

On the other hand, the indeterministic approach does not rely on choices that could prove irreversible—at least in the short-to-medium term, such as artificially changing species or provenance—if the future proves different from that anticipated. However, introducing exotic species may bring new risks to the forests, e.g., new pests and pathogens, or unknown outcomes from competition between tree species.

Furthermore, increasing adaptability to an uncertain future climate could also mean increasing the overall adaptive capacity to react to (unexpected) changes in other aspects, such as market or social expectations. This could be seen as insurance, even by owners who aim at maximizing returns from forest production.

References

- Forest-conservation, whose-forests: toppr.com, Retrieved 11 June, 2020

- Streck, C.; S.M. Scholz (2006). "The role of forests in global climate change: whence we come and where we go". International Affairs. 82 (5): 861–879. doi:10.1111/j.1468-2346.2006.00575.x
- Stamets, Paul (2005). Mycelium running: how mushrooms can help save the world. Ten Speed Press. ISBN 1-58008-579-2
- Wildlife-forestry, global-perspectives-on-sustainable-forest-management: intechopen.com, Retrieved 16 July, 2020
- "Evans, K., De Jong, W., and Cronkleton, P. (2008) "Future Scenarios as a Tool for Collaboration in Forest Communities". "S.A.P.I.EN.S." ""1"" (2)". Sapiens.revues.org. 1 October 2008. Retrieved 30 November 2011

Diverse Aspects of Agroforestry

5 CHAPTER

There are diverse aspects that are studied under agroforestry including modeling agroforestry systems, bamboo based agroforestry and agroforestry watershed management. This chapter sheds light on these diverse aspects of agroforestry for an in-depth understanding of the subject.

MODELLING AGROFORESTRY SYSTEMS

Models are tools that have been developed to make research, management or teaching more effective. In a conventional toolbox you may have many tools such as a hammer, a screwdriver and a saw. In a similar way, a wide range of agroforestry modelling tools is available. The most appropriate agroforestry model will depend on the problem to be addressed and the first step in developing or using a model to examine an agroforestry system is to clearly define your objective. In the same way that you should not use a screwdriver to drive in a nail, or a hammer to remove a screw, using agroforestry models for situations that they were not designed for can cause problems.

A review across temperate and tropical areas suggests that there are six major groups of agroforestry model. These are as follows:

- Allometric or regression models.
- Models where tree and crop yields are included as inputs that describe the environmental impacts of agroforestry.
- Plot-scale models where the focus is on tree and crop growth.
- Architecture models.
- Farm-scale management decision models.
- Landscape models.

Each of these model types are considered in turn, with the greatest focus given to plot-scale models.

Table: Peer-reviewed papers describing the use of allometric, non-growth models, plotscale models of tree and crop growth, architectural models, farm decision models and landscape agroforestry models.

Model Type	Name or Description of Agroforestry Mode
Allometric models	Models relating tree height, diameter, and carbon storage.
Non-growth models of soil carbon	CO_2FIX. SCUAF (Soil Changes Under AgroForestry) model. ICBM/N model.
Plot-scale models of tree and crop growth	ALWAYS (Alternative Land-use with AgroforestrY Systems) silvo-pastoral model. APSim Next Generation. Coffee agroforestry model. Hi-sAFe or STICS[2]. HyPAR: The "Hybrid" tree model with PARCH crop model. WalNulCAS. Yield-SAFE.
Architectural models	"AMAP" canopy methodology. ARCHIMED coconut model. FracRoot root model.
Farm Decision Models	Conceptual model of farm decisions. Knowledge based system to identify agroforestry grants. Model ranking agroforestry threats. BEAM and POPMOD Bio-economic models. ARBUSTRA. Agroforestry Estate Model. Agroforestry Calculator. Forage-SAFE. Financial model including "real options". Welfare maximisation model. Agroforestry adoption models.
Landscape	Landscape models.

Allometry is the study of how the characteristics of a living organism change with size. Allometric relationships are sometimes referred to as 'empirical' models; the Oxford English Dictionary defines empirical as 'based on observation or experience rather than theory or pure logic'. Whilst empirical models are useful for summarising past data and for interpolation, they typically 'contain no information beyond the original data'.

Commercial forestry makes wide use of allometric relationships such as yield tables describing, for example, how tree volume changes with tree height and tree diameter for a specified tree density. For example, in the United Kingdom, there are timber yield data for poplar given at 5-year intervals for a given yield class. It is also possible to use these data to construct further polynomial regression models that allow the interpolation of timber yields for the intervening years. Although the coefficients in such polynomial curves have little physiological meaning, there are examples where there is a physiological basis for a regression. For example the crosssectional area of a tree trunk is a square function of its diameter. Van Noordwijk and Mulia also explain how the fractal nature of some relationships means that it is also possible to develop functional branch analysis models to predict, for example, the weight of branches and leaves, from the stem diameter.

Allometric relationships for agroforestry systems are sometimes reported in scientific papers. They are also used in some detailed growth models. Even in models that calculate the increase in tree dry matter, allometric models can be used to relate changes in tree dry weights to height, diameter and timber volume.

There are a number of agroforestry models that do not calculate plant growth, but can still be used to examine the effect of forestry, agricultural or agroforestry systems on environmental impacts such as changes in soil carbon, soil nutrients or water flow. One example is the Soil Changes under AgroForestry (SCUAF) model which has an annual time step. The user enters the proportion of the area allocated to trees and agriculture in year one and the growth rate of the trees and crops in each year. The model can then be used to estimate the effects of the trees on changes in soil properties including soil erosion, soil organic matter and nutrients, and also on subsequent rates of plant growth. In a similar way, the CO2FIX model has been used to determine the carbon sequestration of poplar agroforestry systems in India both above and below ground. With this model, the user inputs estimates of the annual growth of the trees and the anticipated crop yields.

The most widely cited agroforestry models are those that can be used to predict the effect of tree and crop interactions over time. Sinclair and Seligman defined a crop model as 'the dynamic simulation of crop growth by numerical integration of constituent processes with the aid of a computer'. Matthews et al. defined a crop model as 'a computer program describing the growth of a crop in relation to the environment operating on a time step an order of magnitude below the length of a growing season, and with the capacity to output variables describing the state of the crop at different points in time'.

Most crop and agroforestry models include a 'mechanistic' or 'explanatory' component in that they attempt to describe behaviour at one level by using an understanding of what occurs at a sub-level. Whisler et al. argue that it is generally unnecessary to build a crop model using an understanding on what happens at more than two levels of hierarchy lower. For example the growth of a crop can generally be determined by the interception of light by a canopy and the uptake of water by a root system, rather than by explaining what is occurring at a cellular level. The anticipated advantage of developing mechanistic models based on sub-level processes is that it should allow the model to be used in a wider range of contexts, which also makes it useful as a research tool. However their ability to match observational data may be no better than an empirical model.

Table: A comparison of the principal differences between empirical and mechanistic models.

Type	Mechanistic or Explanatory Models	Empirical or Correlative Models
Scope	Wide	Restricted
How does the model deal with time?	Dynamic	May be static
Connections to observational data	Can be tenuous	Generally good
Accuracy of predictions	Variable	Generally good
Typical use	Research and teaching	Management
Typical complexity	Complex	Simple

Despite the division between mechanistic and empirical models, in practice there is a continuum between pure mechanistic models and pure empirical models. Boote et al. argue that most crop simulation models are a 'mixture of empiricism and mechanism', and Whisler et al. note that all models become empirical at some level. However ideally these empirical relationships occur at a sub-level and, where possible, they are formulated in a way that has a theoretical basis. Another unobserved area in which model use depends on correlation is in the assumption that the climate variables measured in standard weather stations are pertinent to the microclimate experienced by trees or crops in a field.

An important feature of agroforestry systems is the interaction that occurs over time between the trees, the crops and the atmospheric and soil environment. For this reason, most agroforestry models are 'dynamic' as they 'predict how a system unfolds with the passage of time'. Once the objective is clear, Connor writes that there are four steps in the development of a dynamic model.

- There is a need to identify equations for the important flows of material and information.
- There is a need to select values for all the parameters.
- There is a need to set initial values for state variables.
- The model is then run to solve the equations at appropriate time intervals.

An important flow parameter may be the flow of water. The use of soil water by the tree and the crop can be expressed by equations with parameters. The initial state variables are likely to include the initial soil water content, and the change in soil water content over future time intervals is then calculated using a series of equations.

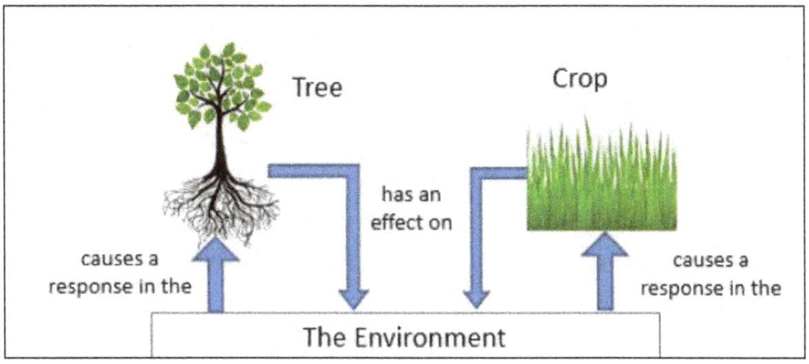

Schematic representation of the interaction between the trees, crops, atmosphere and soil environment that requires 'dynamic' models.

There are arguments in favour of both 'simple' and 'complicated' models. However to paraphrase Albert Einstein, a common approach is that 'things should be as simple as possible, but no simpler'. Box also argued positively for the use of parsimonious models and observed that 'all models are wrong, but some are useful'. Passioura also favoured simple crop simulation models with few parameters, as they were more 'testable'. The minimum number of parameters may increase over time as independent measures of those parameters become available and a model is used to address more specific questions.

Leinweber defined model complexity as a simple count of the number of arithmetic operations a model performs on input data, and he specified two sources of modelling error. One source relates to uncertainties in the input data. If the potential error of these is known, then the errors can combine in a way that can be predicted mathematically. Leinweber assumed that as the number of input parameters increased the margin for error also increased. In addition as the model becomes more complex, it can also become more difficult to describe the flows between spatial components, for example, how surplus of water in one part of the soil will be redistributed to other parts of the soil. The second form of error is a specification error where the model does not accurately describe what happens in practice. Leinweber argues that a model should only be made more complex if it is made more accurate. The assumption is that the specification error should decrease as the complexity increases. The net effect, is that there may be an optimum level of model complexity that minimises the total error due to simplistic assumptions and compound errors in the input data.

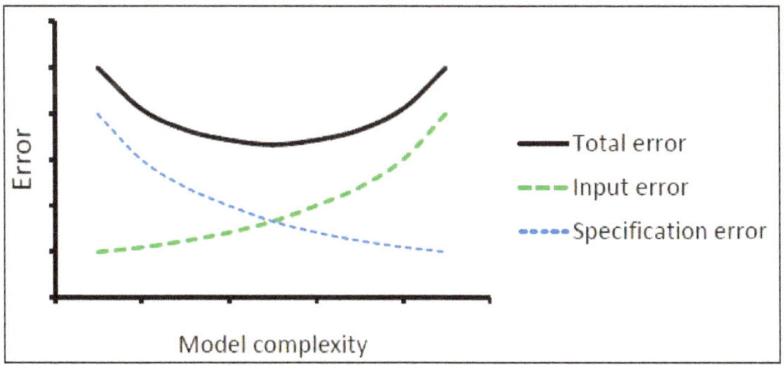

Relationship between model complexity and total error in the upscaling process.

Mechanistic models of tree and crop growth typically build on the physical and biological principles such as those described by de Wit in the Netherlands and Monteith in the United Kingdom. One of the first 'process-based' plant models called ELCROS was described by Brouwer and de Wit. Jackson and Palmer described the results of a light interception model for different shapes of hedgerow orchards demonstrating the changes in diffuse and direct light interception with tree shape, alley width, season and latitude. Charles-Edwards and Thorpe also completed a similar analysis for a hedgerow apple orchard. Jackson and Palmer also described a diurnal light interception model to simulate light interception within an apple intercropping experiment.

One of the simplest mechanistic agroforestry models is Yield-SAFE which includes simple light capture algorithms and was originally designed to describe an alley cropping silvoarable system. As the time step for the model is a day, it is assumed that each part of the alley cropping field will receive some shade from the trees. The incoming daily solar radiation is assumed to be S (unit: MJ m^{-2} day^{-1}). The proportion of the radiation intercepted by trees (f_t) is then defined in terms of the leaf area index of the trees (LAI_t; unit: m^2 tree leaf area per m^2 plot) and a light extinction coefficient (k_t) $f_t = 1 - e^{-k_t \cdot LAI_t}$. Subsequent work showed that it was necessary to change the value of k_t as the trees grew.

$$f_t = 1 - e^{-k_t \cdot LAI_t}$$

The potential radiation available below the trees (S_c) can then be defined using the following equation:

$$S_c = S\,(1 - f_t)$$

Within an alley cropping system it is assumed that silvoarable field is divided into a proportion that is cropped (A_c) and a proportion that is not cropped ($1-A_c$). The proportion of S_c that is intercepted by the understorey crop can therefore be derived, where k_c is the radiation extinction coefficient of the crop and LAI_c is the leaf area index of the crop (m^2 crop leaf area per m^2 of cropped area).

$$f_c = A_c\,(1 - e^{-k_c \cdot LAI_c})$$

Whereas Yield-SAFE partitions the under-storey into a crop and uncropped area, an alternative approach is to partition the agroforestry system in a tree-covered and non-tree covered area. An example of this approach is described by Bergez et al. in the ALWAYS silvopastoral model which used photosynthetically active radiation (PAR). Building on the model of Jackson and Palmer, the model partitioned the system into a proportion covered by a vertical projection of the tree canopy (A_t) and a proportion that was not ($1-A_t$). The mean daily PAR on the pasture (PAR_p) is then calculated as the sum of the sum of the diffuse radiation ($PAR_{diffuse}$) and the transmitted direct radiation ($PAR_{t,direct}$) within the tree canopy and the direct radiation (PAR_{direct}) for the remaining area.

$$PAR_p = (PAR_{diffuse} + PAR_{t,direct})\,At + PAR_{direct}\,(1 - A_t)$$

The value of $PAR_{t,direct}$ was derived from the eqution below where L_d is the leaf area density in the crown and s is the height of the crown.

$$PAR_{t,direct} = PAR_{diffuse}\, e^{(-0.5\,.\,L_d\,.\,s)}$$

The above five equations are examples of how simple agroforestry models can address the interception of solar radiation by a tree and a crop. Models such as Yield-SAFE and ALWAYS can also deal with water competition between the trees and under-storey crops. Another example of a relatively simple agroforestry model is the coffee agroforestry model developed in MATLAB/Simulink based on the BASic FORest simulator (BASFOR) tree model.

A limitation of the Yield-SAFE model is that it is only a one-dimensional model with some consideration of the horizontal extent of the arable crop. An example of a more complex two-dimensional model is the Water, Nutrient and Light Capture in Agroforestry Systems (WaNuLCAS) model as described by van Noordwijk and Lusiana. The original version of the model included a four-layer soil profile (vertical), with four spatial zones (horizontal), and calculation of the water and nitrogen balance and uptake by a crop and a tree. The model used a 'Stella Research' modelling shell with data inputs and outputs provided in Excel spreadsheets.

In an alley cropping context, whilst the two-dimensional WaNuLCAS model can be used to determine the effect of distance from a tree on crop yields, this particular version of the model assumes that light interception in a West-East direction will be the same as in a North-South alignment. Considerations of row alignment require a three-dimensional model. An early three-dimensional agroforestry model combined a detailed tree model called Hybrid with a crop model called PARCH. A more recently developed example of a three-dimensional agroforestry model is Hi-sAFe which combines a tree model with the suite of STICS crop models. The Hi-sAFe model has a switchable toric symmetry module that replicates the simulated scene, allowing it to simulate isolated trees (toric symmetry fully off), tree hedges (toric symmetry off in one direction), alley cropping agroforestry (full toric symmetry) and forest edges (toric symmetry off in y and half-x axis), as well as crop and forest monocropping. In Hi-sAFe, root competition is addressed using a three-dimensional opportunistic tree root growth module based on a cellular automata algorithm. This allows the simulation of tree root systems that can adapt continuously to the soil environment as modified by the weather and the crop competition. Tree root pruning, tree branch pruning, tree stand thinning and a large panel of crop management schemes are also available. Another feature of Hi-sAFe is the possibility of considering the impact of a fluctuating seasonal water table on tree-crop competition.

The initial use of the Hi-sAFe model included an examination of the clumpiness of leaves in trees and the simulation of walnut-wheat alley cropping for the duration of the tree life cycle. More recently Dupraz et al. describe the use of the Hi-sAFe model to determine the best tree row orientations to maximise the systems outputs at various latitudes, from the tropics to boreal regions. This analysis showed that in terms of maximising understorey crop yields and uniformity, the best tree line orientation was East to West at low latitudes and North to South at high latitudes. Recent developments include the use of the model to assess the resilience of agroforestry systems to climate change, the nitrogen leaching control by the trees and the long-term C sequestration in the soils.

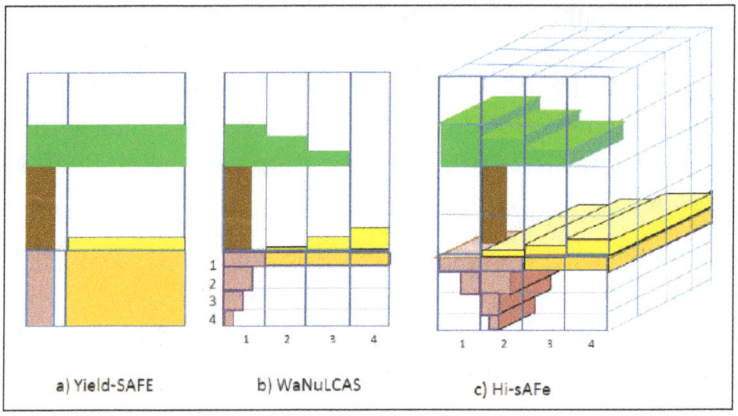

(a) The Yield-SAFE model comprises of two above-ground zones (crop and no crop) and one below-ground zone (soil depth). (b) The WaNuLCAS model described by van Noordwijk and Lusiana comprises four root zone depths and four distances from the tree. (c) Hi-sAFe is an example of a three-dimensional agroforestry model.

One potential advantage of the Hi-sAFe model is that the crop component is based on an established crop model called STICS, which is supported by a strong user community. Results reported by Artru et al. have demonstrated how different shading patterns affect the yields predicted using the STICS model. Luedeling et al. also describe some of the benefits in terms of flexibility in modelling agroforestry systems using the Agricultural Production Systems Simulator (APSIM) framework. The APSIM framework was initiated in 1994 and a 'next generation' of APSIM was released in October 2014. Smethurst et al. describe the use of APSIM Next Generation framework to provide a two-dimensional (2D) representation of Gliricidia-maize intercropping, which is reported to describe accurately crop yield and water and soil carbon dynamics.

Architectural models of tree growth are the fourth type of agroforestry model. De Reffye et al. describe the development of a canopy architecture model called AMAP-mod by CIRAD in France, where tree architecture is described in terms of the form of 'growth units'. Models such as the 'FracRoot' model have also been used to describe below-ground root architecture. Lamanda et al. describe the use of threedimensional architectural models to assess light availability and root bulkiness in coconut agroforestry systems in Vanuatu in the South Pacific.

The fifth group of agroforestry models comprises models where the primary focus is to improve decision-making by farmers and others. Some comprise conceptual models, but many assume that profitability is an important determinant of whether a farmer will adopt a system.

Sinclair and Walker describe a method to model indigenous ecological knowledge on agroforestry. They demonstrate that formal representation of such models can increase the coherence, consistency and completeness of our understanding of agroforestry systems. Meylan et al. also describe the development of a conceptual model to identify the constraints and trade-offs associated with four types of farmers practising agroforestry with coffee production. The model includes environmental factors (e.g. climate), management factors, agronomic and hydrological processes, and outputs such as yield, gross margin and soil erosion losses. The process results in a series of system-flow diagrams showing the relationships between the various factors, processes and outputs. They argue that this can serve as a preliminary stage before attempting to identify the key points of potential interventions in the system. In a similar way, flow diagrams were developed by Acuña et al. in their development of a knowledge-based system model to create decision trees related to agroforestry grants in Spain. Moral et al. in Spain also examine the rationale for on-farm decisions drawing on a survey of dehesa management based on Strength, Weaknesses, Opportunities and Threats (SWOT) model. The survey identified 68 potential threats to the system, and a Rasch ranking methodology was used to develop a model that describes the most important threats.

It is possible to undertake a financial cost-benefit analysis of agroforestry systems without a detailed biophysical model of tree and crop yields. For example Dube et al. describe the financial benefits and costs of a eucalyptus agroforestry system in Brazil. However in their review of financial models of silvoarable systems, Graves et al. found five models that could incorporate yield and financial data. The models included WaNuLCAS with a financial component, POPMOD, ARBUSTRA, the Agroforestry Estate Model and the Agroforestry Calculator. Graves et al. found that whilst ARBUSTRA and the Agroforestry Calculator relied on direct inputs of yield data, WaNuLCAS, POPMOD and the Agroforestry Estate Model contained biophysical models of differing complexities that were used to generate the yield data for the financial analyses. For WaNuLCAS and Agroforestry Estate Model, this information flowed in one direction from the biophysical component of the model to the financial component of the model. In the POPMOD model, it was possible for management feedback from the financial component of the model, to link back to the biophysical component and modify the yields obtained.

The Farm-SAFE model developed during the SAFE project is an example of a financial model, developed to evaluate the profitability, practicality and feasibility of silvoarable systems from a farmer perspective. It used annual tree and crop yield data on arable, forestry and agroforestry systems, derived from the Yield-SAFE model, which was then used to compare the financial benefits and costs. As it is a farm-scale model, it can be used to model different arable, forestry and agroforestry systems across a farm reflecting, for example, contrasting levels of soil fertility. Another example of a recent model combining financial and yield data for grass-oak tree systems in Spain and Portugal is Forage-SAFE. This uses a relationship for grass growth at varying distances from an oak tree to calculate what the optimal stocking rate should be for livestock at different tree densities and grass productivity levels.

One of the important points in completing a useful financial analysis of an agroforestry system is the ability to complete analogous analyses of 'counterfactual' forestry and agricultural systems. Thus, for example, Farm-SAFE, ARBUSTRA and POPMOD provide a financial assessment of arable and forestry systems against which the financial performance of agroforestry can be compared. One effect of this is that most agroforestry financial models are effectively combined agricultural, agroforestry and forestry models, for example, they can compare a wide range of land uses. In order to account for the impact of the different systems on labour and machinery costs, it is generally accepted that the comparison should be based on net margins, rather than gross margins. Freya et al. also describes how the impact of changes in price, after the establishment of an agroforestry or forestry system, can affect management choice using a stochastic and 'real options' model. Farm-SAFE, for example, contains the option of changing future grants, prices and costs either in a continuous phased or discontinuous one-off pattern from a specified year in the future.

Whilst financial models can be used to compare the attractiveness of an agroforestry system with arable or forestry systems for a farmer, the attractiveness of agroforestry from a societal perspective requires consideration of wider societal costs and benefits. Possible benefits may be increased carbon sequestration or reduced nitrate leaching which does not have a readily available market value. Through the use of cost-benefit analysis and environmental valuation data, it is possible to impute values for many of the outputs that do not have immediate market value. Since 2014, Farm-SAFE has been developed to incorporate non-market societal costs and benefits of arable, forestry and agroforestry systems by incorporating life cycle assessment and valuation data. Life cycle assessment data can be linked with management decisions to quantify the environmental burdens associated with tree and crop production in the modelled arable, forestry and agroforestry systems. This is then linked with valuation data derived using a benefits-transfer approach in order to quantify the non-monetary value of changes in carbon sequestration, and erosion, nitrate and phosphorus emissions to water. Dhakal et al. also describe the use of a constrained welfare maximisation model to identify the opportunity of farmers implementing agroforestry in Nepal. The analysis was based on a household income model, a household production model and a community welfare model.

The adoption of agroforestry is not solely determined by a financial and economic analysis. Lovrić et al. describe the use of an analytic network process to quantify the driving forces between agroforestry uptake in Mediterranean Europe in environmental, economic and social areas.

A common question in government-sponsored research is to determine the effect of the widespread uptake of agroforestry. This can take place at a landscape, regional or national scale. Models such as ARBUSTRA and Farm-SAFE can be used as landscape models since they are able to model a range of different arable, forestry and agroforestry systems under different environmental and management conditions. The results can then be aggregated to the landscape scale by summing the outputs of the different systems. The biophysical assessments are typically based by running versions of the biophysical, financial and economic models described above within a geographical information system (GIS) framework. Such an approach was used in the SAFE project by Palma et al. In the AGFORWARD project, a similar approach was used by Kay et al. Another example of a landscape model linking GIS with other models is Polyscape. The Polyscape model uses algorithms to determine the effect of land cover changes, such as agroforestry, on flooding, soil erosion, carbon storage and agricultural production. In China, Zeng et al. describe the use of GIS with a land-water-environment model to determine the effect of afforestation or increased agricultural production. It is interesting that they refer to their model as an agroforestry ecosystem model – whereas in reality it is a model that can address forestry, agroforestry and agriculture.

MEDICINAL AND AROMATIC PLANTS IN AGROFORESTRY SYSTEMS

Many plants in traditional agricultural systems in the tropics have medicinal value. These can be found (either planted or carefully tended natural regenerations) in homegardens, as scattered trees in croplands and grazing lands, and on field bunds. Many Acacia species found in Africa such as Acacia nilotica, A. seyal, A. senegal and A. polyacantha, as well as several species in African croplands (e.g. Faidherbia albida, Vitellaria paradoxa, Adansonia digitata, Markhamia lutea and Melia volkensii) have medicinal value. Similarly, 'arjun' (Terminalia arjuna) in India, chinaberry (Melia azederach) in Asia and Erythrina species in Latin and Central America combine many uses including medicinal. Holy Basil or 'tulsi' (Ocimum sanctum), drumstick (Moringa oleifera) and curry leaf (Murraya koenigii) are backyard plants in many Indian households and they are routinely used for common ailments or in food preparations. A number of plants used as live fences around home compounds such as henna (Lawsonia inermis) in India, Ipomoea carnea ssp. fistulosa in Bolivia and Asian countries and Euphorbia tirucalli around crop fields in Africa have medicinal values. Although the medicinal value of these plants is 'exploited' locally, they are seldom used for commercial purposes (except in the case of commercially exploited neem). In fact, many of these species are valued for poles, fuel wood, fodder, fruit, shade and boundary demarcation and their medicinal value is secondary.

Table: Examples of tree species reported to have medicinal values grown in traditional agroforestry systems in the tropics.

Tree Species (Common Name)	Agroforestry System	Medicinal Use
Acacia nilotica (babul)	Field bunds, scattered trees in croplands, woodlots and grazing lands in East Africa and Indian sub-continent.	Gum used for treating diarrhea, dysentery, diabetes, sore throat, bark used to arrest external bleeding.

Azadirachta indica (neem)	Woodlots, scattered trees, shelterbelts in Africa and India.	Digestive disorders, malaria, fever, hemorrhoids, hepatitis, measles, syphilis, boils, burns, snakebite, rheumatism.
Erythrina spp.	Shade tree in coffee, live fence.	Different Erythrina species have different uses.
Parkia biglobosa (ner´e or locust bean tree)	Parklands in West Africa.	Piles, malaria, stomach disorders, jaundice.
Prosopis cineraria (khejri)	Scattered trees in croplands in arid to semiarid areas mostly in northwestern India.	Flowers for blood purification and curing skin diseases. Bark against summer boils, leprosy, dysentery, bronchitis, asthma, leucoderma and piles.
Tamarindus indica (tamarind)	Field bunds and scattered trees in croplands in semiarid India and Africa.	Fruit pulp is used in Indian medicine as refrigerant, carminative and laxative. It is also recommended in febrile diseases and bilious disorders.

Forests and Forest Plantations

MAPs growing in forests require (or tolerate) partial shade, moist soils high in organic matter, high relative humidity, and mild temperatures. Cultivation of such MAPs can be taken up in thinned forests and cleared forest patches, and as intercrops in new forest plantations. In China, cultivation of medicinal plants has been an age-old practice under the name of 'silvo-medicinal' systems. In northeast China, ginseng (Panax ginseng) and other medicinal plants are grown in pine (Pinus spp.) and spruce (Picea spp.) forests; in central China, many medicinal plants are planted with Paulownia tomentosa and in southern China medicinal herbs are often planted in bamboo (Bambusa spp.) and Chinese fir (Cunninghamia lanceolata) forests. In Yunnan province, China, traditional 'Dai and Jinuo' agroforestry systems involve the medicinal crop Amomum villosum in the forest areas cleared of undergrowth. The forest is thinned to give 30% to 40% shade and seedlings or cuttings are planted, which produce an average dried fruit yield of 375 kg ha^{-1} per annum. Gupta listed a number of indigenous understory herbs and shrubs that can be produced as part of forest farming or in new forest plantations to improve economic returns from forests in India.

A farmers' cooperative in the northern lowlands of Costa Rica has successfully demonstrated cultivation of the medicinal herb 'raicilla' (Cephaelis ipecacuanha) in natural forests for export to the Netherlands and Germany. American ginseng (Panax quinquefolium), a medicinal herb exported to China from the United States and Canada is grown as an understorey in red maple (Acer rubrum) forests or deciduous hardwoods such as black walnut (Juglans nigra) and sugar maple (Acer saccharum), instead of growing under artificial shade with considerable expense. Indeed, cultivation of ginseng and several other medicinal plants in the forests is a common and growing form of forest-farming practice of agroforestry in North America. Light demanding understory species (e.g., Echinacea sp.) may be intercropped initially to provide early returns from plantations and after canopy closure, shade-tolerant species such as ginseng and goldenseal can be intercropped. Studies in New Zealand have indicated that the American ginseng can be successfully grown under Pinus radiata with best growth under a tree stand of 130 stems ha^{-1}. In addition to providing shade, the trees may also benefit the understory component from hydraulically lifted water. Fungal diseases are a major concern in forest farming but the application of fungicide can be detrimental to the forests' health, therefore proper spacing and mixed cropping is recommended. Mechanical cultivation may not be feasible under forested conditions so labor availability needs to be considered as a constraining factor.

As in the taungya system, newly established forest plantations can be intercropped with MAPs similar to food crops until the trees cover the ground. The participation of the local people with right to share benefits of the plantations, especially ownership to crops, has helped governments to establish and protect large-scale tree plantations without conflict with the local people in many Asian countries. The same approach can be employed for the cultivation of MAPs in the new plantations. In the rehabilitation of degraded forestlands, participatory planning and implementation with local communities and economic benefits from an early stage onwards will ensure commitment of the people.

The intensity of shade experienced by the understory MAPs growing in forests and tree plantations affects their growth and chemical composition. Growth, and bark and quinine yields of Cinchona ledgeriana grown on the Darjeeling hills, India, increased when it was associated with shade of five species compared with that of a nonshaded stand. The best yields were obtained when C. ledgeriana was planted under the shade of Crotalaria anagyroides or Tephrosia candida initially and Alnus nepalensis in the later stages. Alnus is planted at 3.6 × 3.6 m and is progressively thinned to 14.4 × 14.4 m as C. ledgeriana grows. The other factors that affect MAPs yields are their growth cycle and nutrient inputs. The optimum rotation for C. ledgeriana in the Himalayan region is 16 years. Dioscorea deltoidea grown in deodar (Cedrus deodara), fir (Abies spp.) and spruce plantations attains exploitable tuber size in about 10 years.

Homegardens

Homegardens are complex agroforestry systems involving many plant species characterized by different morphology, stature, biological function and utility, practiced mostly in the humid and subhumid tropics. Food, fruit and timber species may dominate the homegardens and occupy the middle and upper strata, but medicinal plants, spices and vegetables occupy the lower stratum. Three categories of medicinal plants could be noted in homegardens: species used exclusively for medicine, horticultural or timber species with complementary medicinal value, and 'weedy' medicinal species. While the first two categories are deliberately planted the latter group is part of spontaneous growth. The species composition, plant density and level of management vary considerably depending on the soil, climate and market opportunity and cultural background of the people. Homegardens of individual holdings generally cover small parcels of land and are established around homesteads. Although these systems in the past were mostly seen to meet the home needs of small-scale farmers in the forest margins, increased urbanization, transport and market opportunities in recent times are helping to produce cash value crops. Multistrata systems involve fewer species (three to 10 species) than in homegardens in definite planting arrangement and can be designed for home consumption as well as commercial production.

Table: Examples of commercially valuable medicinal plants under cultivation or that can be produced as understory components in forests and tree plantations.

Name	Common Name	Plant Type	Parts Used	Medicinal Use	Location
Aconitum heterophyllum	Atis	Tall herb	Rhizomes	Hysteria, throat diseases, astringent	Alpine and sub-alpine Himalayas
Amomum subulatum	Large cardamom	Perennial herb	Seeds	Stimulant, indigestion, vomiting, rectal diseases	Sub-Himalayan range, Nepal, Bhutan
Amomum villosum	Saren	Perennial herb	Seeds	Gastric and digestive disorders	China
Caulophyllum thalictroides	Blue cohosh	Perennial herb	Roots	Gynecological problems, bronchitis	North America
Cimicifuga racemosa	Black cohosh	Perennial herb	Roots	Menses related problems	North America
Chlorophytum borivilianum	Safed musli	Annual herb	Tubers	Male impotency, general weakness	India
Costus speciosus	Crepe ginger	Cane	Leaves, stem, rhizomes	Purgative, depurative and as a tonic	India
Dioscorea deltoidea	Himalayan yam	Vine	Tubers	Source of saponins and steroids	India, Pakistan
Echinacea purpurea	Coneflower	Perennial herb	Roots, rhizomes	Enhancing immune system	North America
Hydrastis canadensis	Goldenseal	Perennial herb	Rhizomes	Tonic	North America
Panax ginseng	Ginseng	Herb	Roots	Tonic	China, Korea, Japan
Panax quinquefolium	American ginseng	Perennial herb	Roots	Tonic	North America
Cephaelis ipecacuanha	Raicilla, Ipecac	Shrub	Roots	Whooping cough, bronchial asthma, amoebic dysentery	Brazil, India, Bangladesh, Indonesia
Rauvolfia serpentina	Rauvolfia	Shrub	Roots	Hypertension and certain forms of insanity	Sub-montane zone, India
Serenoa repens	Saw palmetto	Shrubby-palm	Fruits	Swelling of prostate gland	Southeastern USA

Medicinal plants are an invariable component of homegardens, whether they are in the Peruvian Amazon, on the slopes of the Mt. Kilimanjaro in Tanzania, or in the humid and semiarid Cuba. The species composition differs depending on cultural background, distance from markets and influence of tourism. Medicinal plants accounted for about 27% of total plant species in the homegardens in Amazon, 56% in northern Catalonia (Iberian Peninsula) and 45% in the floodplains of the river Jamuna in Bangladesh. In the Soqotra Island, Yemen, endemic medicinal plants such as Aloe perryi, Jatropha unicostata and Commiphora ornifolia are cultivated in homegardens. On St. Croix and St. Thomas, U.S. Virgin Islands, the medicinal trees neem, moringa, and noni (Morinda citrifolia) have become popular in Homegardens.

The slash-and-burn system, which is so widely practiced in the humid and sub-humid tropics, has reduced soil fertility, crop yields, biodiversity and mature forest vegetation. Agroforests that mimic local ecosystem processes can be developed to provide livelihoods for farmers while protecting and preserving forest reserves and biodiversity. Based on the analysis of plant species in different strata and analog hypothesis, an agroecological system can be designed that meets the farmers' needs of food, timber, medicinal plants and other NTFPs, efficiently uses the on-site resources and protects the natural resource base of the region. Integration of medicinal plants in agroforests and multistrata systems provides an answer to the plantation commodity conundrum. These complex agroforestry systems can be utilized to grow MAP species for home use and markets. MAPs produced in agroforests may be targeted to niche markets to secure higher premium on the premise of better quality similar to those harvested from wild. The forest-type environment of these systems facilitates the integration of species that generally grow in the forest and thereby helps conserve the endangered species and produce them for markets. Homegardens and multistrata systems are recognized as the most productive, remunerative, environmentally sound and ecologically sustainable alternative land use systems to slash-and-burn systems and 'alang alang' (Imperata cylindrica)-infested degraded lands in humid tropics. High value MAPs may be integrated in the newly establishing homegardens and multistrata systems.

Riparian Buffer Zones

An agroforestry system that has received considerable attention in North America is riparian buffers zones. Riparian buffer zones can improve water quality and protect streams and rivers from degradation by nutrient loading and chemical pollutants from agriculture and urban areas, from erosion by attenuating peak flows and provide habitat for wildlife. NTFP production can help defray the cost of buffer zone installation and maintenance. Slippery elm (Ulmus rubra), harvested for its aromatic and medicinal inner bark, is commonly found in riparian areas in North America. Riparian buffer zones are an ideal location for the production of this species, which suffers from commercial over-exploitation and the Dutch elm disease.

Intercropping of MAPs

Two types of intercropping systems can be distinguished involving MAPs:

- Medicinal plants as upperstory trees.
- MAPs as intercrops in other tree crops.

Medicinal plants as overstory trees: Coffee (Coffea arabica), cacao (Theobroma cacao) and tea (Camellia sinensis) are traditionally grown under shade offered by multipurpose trees that produce timber, fruit, flowers, nuts, palms etc. Medicinal tree species that grow tall and develop open crown at the top can also be used for this purpose, for example yongchak (Parkia roxburghii) in India, the protein rich seeds of which are used to treat stomach disorders and Ginkgo biloba in China, the nuts of which are used in Chinese medicine and fetch high value. In Ivory Coast, 19 of the 41 tree species planted as shade trees in coffee and cacao provide pharmaceutical products for traditional medicine. New plantations of coffee, tea, and cacao offer scope for cultivation of forest medicinal trees that are under demand. However, research needs to identify the medicinal trees that can be grown in association with these plantation crops and develop management practices for them. Tall and perennial medicinal trees that need to be planted at wider spacing such as Prunus africana, Eucalyptus globulus (for oil), sandalwood (Santalum album), ashok (Saraca indica), bael (Aegle marmelos), custard apple (Anona squamosa), amla (Emblica officinalis), drumstick or moringa (Moringa oleifera) and soapnut tree (Sapindus mukorossi) can be intercropped with annual crops in the early years until the tree canopy covers the ground. Some of the medicinal trees may allow intercropping for many years or on a permanent basis depending on the spacing and nature of the trees. The intercrops give some income to farmers during the period when the main trees have not started production.

Medicinal plants as intercrops: Many tropical MAPs are well adapted to partial shading, moist soil, high relative humidity and mild temperatures, allowing them to be intercropped with timber and fuel wood plantations, fruit trees and plantation crops. Some well-known medicinal plants that have been successfully intercropped with fuel wood trees (e.g., Acacia auriculiformis, Albizia lebbeck, Eucalyptus tereticornis, Gmelina arborea, and Leucaeana leucocephala) in India, include safed musli (Chlorophytum borivilianum), rauvolfia (Rauvolfia serpentina), turmeric (Curcuma longa), wild turmeric (C. aromatica), Curculigo orchioides, and ginger (Zingiber officinale). Only 10 out of 64 herbaceous medicinal plants tried in intercropping with two-year old poplar (Populus deltoides) spaced 5 m apart gave poor performance, indicating that many medicinal plants can be grown in agroforestry systems. The trees may benefit from the inputs and management given to the intercrops. Short stature and short cycle MAPs and culinary herbs are particularly suited for short-term intercropping during the juvenile phase of trees. Wherever markets are established, MAPs are remunerative alternative intercrops to the traditionally grown annual crops. The number of years MAPs can be intercropped with a given tree species depends on the size and intensity of its canopy shade, tree spacing and management, especially pruning of branches and nature of the MAPs. Shade-tolerant and rhizomatic MAPs can be grown on a longer-term basis in widely spaced plantations.

Intercropping of medicinal plants in coconut (Cocos nucifera) and arecanut (Areca catechu) stands is an age-old practice in India and other parts of south and Southeast Asia. These palms allow 30% to 50% of incident light to the underneath, which is ideal for some MAPs, including cardamom (Elettaria cardamomum). Kacholam or galang (Kaempferia galanga) – a medicinal herb. Kacholam intercropped in a 30 year-old coconut plantation produced 6.1 Mg ha^{-1} of rhizomes compared with 4.8 Mg ha^{-1} as a sole crop. Twelveyear old coconut trees did not adversely affect the growth and yields of a number of medicinal species grown as intercrops compared to the yields in the open. Some of these intercrops may cause small reduction in arecanut yields but the combined returns from both the components are greater than from arecanut alone. Another plantation crop intercropped with MAPs is rubber (Hevea brasiliensis), for example with Dioscorea floribunda in the state of Assam in India and with Amomum villosum in Yunnan province of China. In Fujian Province, China, Cunninghamia lanceolata – an important timber tree – is intercropped with a variety of cereals, cash and medicinal and oil-producing crops. Many of the medicinal herbs commonly grown in thinned forests can also be grown intercropped with trees.

In the Caribbean islands, there has been increased interest on alternative crops that have better economic potential than traditional crops. For example, in the U.S. Virgin Islands, a number of farmers are now opting for specialty crops such as the West Indian hot peppers (Capsicum chinense), thyme (Thymus vulgaris) and chives (Allium schoenosprasum) instead of vegetables. The prospects of growing indigenous MAPs such as 'japana' (Eupatorium triplinerve), worrywine (Stachytarpheta jamaicensis), inflammation bush (Verbersina alata) and lemongrass (Cymbopogon citratus) in association with the medicinal trees noni (Morinda citrifolia) and moringa have been explored at the University of the Virgin Islands, St. Croix. These local herbs are commonly used as bush teas and very popular in the Caribbean. Medicinal plants and herbs in intercropping produced similar yields to those in sole cropping at the first harvest, but they tended to be lower than in sole cropping at subsequent harvests.

Table: Important medicinal trees currently exploited from Peruvian Amazon forest that can be grown in multistrata or homegarden agroforestry systems.

Name	Plant Type	Part Used	Medicinal Use
Cedrella odorata	Tree	Bark	Snake bite.
Croton lechleri	Tree	Latex	Swellings.
Euterpe precatoria	Palm	Roots	Kindness, diabetes.
Ficus insidpida	Tree	Latex	Anemia.
Maytenus macrocarpa	Tree	Bark	Arthritis, diarrhea, stomach disorders.
Tabebuia serratifolia	Tree	Bark	Vaginal diseases, hepatitis, diabetes, arthritis
Uncaria tomentosa	Shrub	Bark	Infections, cancer.

Aromatic Plants as Intercrops

Studies have been conducted in India since mid-1980s on the feasibility and economic aspects of intercropping aromatic plants with timber trees. Experiments at different sites in India demonstrated that aromatic Mentha spp. (M. arvensis, M. piperita, M. citrata var. citrata, M. spicata, M. cardiaca and M. gracilis) and Cymbopogon spp. [lem-on- grass (C. flexuosus); Java citronella (C. winterianus); and palmarosa (C. martinii)] can be grown intercropped with Populus deltoides or Eucalyptus spp. for three to five years after planting the trees. Some of the above aromatic species and Pelargonium sp. (geranium) can also be intercropped with another essential-oil-yielding tree Eucalyptus citriodora (lemon- or citron-scented gum). Tree growth was generally little affected or improved probably due to the inputs given to the intercrops. Herb growth and oil yields of these aromatic plants were not affected during the first two years of intercropping, but they decreased slightly from the third year compared with sole crops. Essential oil profiles of citronella and palmarosa grown in open and partial shade were not significantly different, but the oil of menthol mint (M. arvensis) produced in the open was richer in menthone while that produced under shade was richer in menthol. This implies the need to consider the effect of intercropping on oil quality of some aromatic plants. These intercropping systems gave two to four times greater economic returns than sole cropping of the components. These studies indicate that aromatic plants can be profitably grown in association with linearly growing fuel- and timber trees such as L. leucocephala, Casuarina spp., and Grevillea robusta.

Table: Yields of aromatic plants in intercropping with trees relative to those in sole cropping at a few locations in India and Nepal.

Location	Tree Species and Spacing	Aromatic Plant	Intercrop Yield as % of Sole Crop		Comments
			Herbage	Oil	
Pantnagar, India	Populus deltoids 5 × 4 m	Cymbopogon flexuosus	82	82	Results averaged over five years.
		C. martinii	96	86	
	Eucalyptus hybrid 2.5 × 2.5 m	C. flexuosus	96	90	
		C. martinii	100	86	
Pantnagar, India	Populus deltoids 5 × 4 m	Mentha arvensis	90	90	Results averaged over three years.
		M. piperita	85	88	
		M. citrata	90	90	
		M. spicata	87	87	
		C. flexuosus	99	97	
		C. martinii	99	98	
		C. winterianus	98	99	

Hyderabad, India	Eucalyptus citriodora 3 × 2 m	C. flexuosus	87	87	Results averaged over two years.
		C. martinii	87	87	
		C. winterianus	112	91	
		Pelargonium sp.	91	90	
Nepal	Dalbergia sissoo 10 × 2 to 8 m	C. winterianus	-	100	Results average over two years.

MAPs in other Agroforestry Systems

Aromatic grasses such as vetiver (Vetiveria zizanioides), lemongrass and citronella (Cymbopogon nardus) can be grown on field bunds and soil conservation bunds in croplands. Vetiver has extensively been tested and is being promoted for planting in contour strips or as a live hedge barrier and to stabilize terrace risers on sloping lands in a number of countries, for example, in India, Fiji, Haiti and Indonesia. In the hill subsistence farming systems, maintaining woody perennials in contour strips across slopes and around fields is a common practice. The trees and shrubs produce fodder and firewood and reduce soil erosion hazards allowing crop production on steep slopes where it would otherwise be difficult. Many locally available medicinal plants can be incorporated into these systems. The harvest of grass strips on sloping lands for economic purpose should be planned carefully to avoid soil erosion. Indiscriminate digging of vetiver for distilling oil from its roots has often resulted in worst erosion problems in Haiti.

Hardy medicinal and fruit trees can be grown in community lands in villages and degraded lands. In the state of Madhya Pradesh, India, amla – a rich source of vitamin C, and custard apple, for both fruit and medicine, have been taken up as part of social forestry. Amla and bael withstand saline and alkali conditions and can be grown on salt-affected soils. Similarly, in West Bengal, India, Costus speciosus, a diosgeninproducing plant, is recommended as a major understorey crop in social forestry programs. Other medicinal trees that can be grown on wasted lands include soapnut tree and wood apple. A study of the medicinal species in the Shinyanga region of northwestern Tanzania indicated that the top 10 priority species can be integrated into land use systems such as ngitili (grazing lands planted with economically valuable trees and shrubs), scattered trees in croplands, on field bunds or in homesteads. It is recognized in India that many medicinal trees can be integrated into croplands by planting them on field bunds or as scattered trees.

ANIMALS IN AGROFORESTRY

Most modern systems of raising animals separate animals from their natural environment. In these systems, many of the cycles of nature which are beneficial to both animals and their environment are broken.

The use of animals can be a controversial topic. In many situations, especially in the tropics, animals such as cattle, goats and sheep are associated with ecological degradation and deforestation. There is no doubt that the inappropriate use of animals and poor management leads to many environmental problems. However, this does not necessarily have to be the case.

If planned and managed properly, appropriate kinds of animals can be key components in sustainable farming systems. Agroforestry systems that incorporate animals with tree crops can enhance important cycles of nature such as nutrient cycling and balancing of insect populations while reducing energy intensive management techniques. The key is to integrate the natural needs, behaviors, and products of animals with the environment provided by the agroforestry system in a way that maximizes the benefits to the animals and to the system as a whole.

As a part of their normal behavior, animals forage for plant foods and insects, spread manure, and dig and scratch in the earth. Many of the products and behaviors of animals, for example manure, are considered problems or pollutants in modern conventional production of e.g. poultry and livestock. However, these animals can perform many beneficial functions when integrated into an agroforestry system.

- Some products and services of animals for agroforestry: grazing/weed maintenance (livestock, poultry).
- Insect control (poultry).
- Cleaning of fallen fruit/nuts, other organic wastes (livestock, poultry).
- Spreading nutrients in the form of their manure (livestock, poultry).
- Scratching and digging to prepare for planting (pigs, chickens, turkeys, etc.).

Similarly, agroforestry systems can have an excess of weedy undergrowth, fallen fruit or nuts and insect populations. Often these excesses are controlled through human intervention--weed cutting, insect control, etc. Animals, when selected carefully for the right agroforestry system can carry out many of the maintenance functions through their normal behavior.

Some products and services of agroforestry systems for animal husbandry:

- Favorable habitat.
- Access to a diverse diet.
- Access to medicinal plants and insects.
- Natural population densities.

In most commercial production systems, chickens are caged in large facilities. Food, water, and grit is provided to them via a massive, energy-intensive infrastructure of farming and transportation. In this unnatural and crowded environment, chickens are often stressed, and producers therefore also supply them with medication from the pharmaceutical industry. The manure from these factory-like facilities is in many cases considered a pollutant, and in some cases is contaminated. The resulting products of eggs or meat are questionable in quality in the eyes of many consumers. (In fact, consumers gaining a greater understanding of the environmental impacts and health conditions of these kinds of animal factories are limiting their intake of animal products.) The chickens are deprived of their natural activities and a healthful environment, and the humans must work very hard to supply the chickens with food and medicine and dispose of their waste. As Mollison asks: "Who is working for whom?"

In an integrated system, by comparison, chickens are raised in managed habitats where they have regular access to a diverse diet rich in plants and insects, as well as their other needs such as shelter and water. In return, the chickens control understory growth through foraging and scratching, eat large quantities of insects, and return the nutrients back to the system through their manure. In such a managed system, the chicken's behaviors and wastes play their natural role in what is a complex natural system. By using the natural cycles, raising chickens in an agroforestry system can reduce human inputs to both the chickens and the plants.

Examples of animals in agroforestry systems:

- Chickens in garden systems, orchards and forest.
- Ducks in aquatic environments, such as rice paddies.
- Geese to control grasses in orchards.
- Livestock in forest plantations (silvopasture).

Of course an animal in the wrong place at the wrong time can create problems, not solve them. For example, the wild pigs prevalent in Hawaii destroy new plantings, damage trees, and spread weedy exotic species throughout native forest. The same pigs have been used successfully for decades in traditional Hawaiian farm forestry in controlled paddocks to graze and consume farm residues, and to provide meat for sustenance. In another example, chickens (and other poultry) are notorious for damaging new plantings, especially gardens. Once an area is prepared and planted, it is essential to exclude the chickens (or other poultry) until the plants can hold their own against them.

Advantages of integrating animals:

- Less reliance on outside inputs.
- Reduced human maintenance.

- Better.

Disadvantages:

- Animals must be carefully selected for particular systems.
- Animals require tending and management of grazing.
- Food and water may need to be supplemented at times.
- Health care required.
- Requires a special set of skills and knowledge.

The Right Animal in the Right Place at the Right Time

Clearly, the introduction and management of animals must be carefully thought out, to ensure that the right kinds of animals are introduced to the right place, and at the right time. Some environments, like savannas, are naturally suited for certain types of grazing animals; in other areas smaller animals and birds might be more appropriate. For some very fragile ecosystems, animals may not be appropriate at all.

Also, the type of animal is important for the environment, and farmers should not feel they have to limit themselves to conventional animals. All over the world, farmers are branching out to incorporate animals that are more appropriate to their local environment, culture, and market demands. Examples of alternative animals being used successfully include iguanas, small forest deer, and ostriches.

Commingling the right animals with the right agroforestry system can create highly productive systems that require less human intervention for maintenance and control. In this way, animal production can be sustainable.

INSECTS AND PESTS IN AGROFORESTRY

The science of agroforestry is of recent origin, although the practice is age old. There are numerous types of agroforestry system in different parts of the world. There have been few studies of insect pests in agroforestry context, although insect pests of crops that are components of agroforestry systems have been studied. Scanty information is available about the insects associated with the multipurpose trees and shrubs that are gaining greater economic importance as components of agroforestry systems. In any particular location, the insect fauna occurring on certain plant species is more or less the same, whether that plant species is in monoculture or in a polycultural assemblage such as an agroforestry system. However, the activities of these insects are not likely to be identical in any two situations. There are several factors that influence the activities of insects in agroforestry.

The mechanisms governing the insect-pest situation in agroforestry systems are yet to be investigated fully, and there have been no studies comparing the insect-pest situation in monoculture or block plantation and in agroforestry combinations. Most publications that mention pests of agroforestry either contain information on the insect pests of one component or underline the need for research in this field. The management of insect pests through agroforestry field design has been the subject of discussion in a few publications. Liping suggested directions for research on biological control of pests and diseases in agroforestry systems.

There have been several studies on insect activities in windbreaks, shelterbelts and hedges under temperate conditions, but these were not in an agroforestry context as most of them were restricted to woody plants. Insect dynamics in windbreaks have been investigated by Lewis, Lewis and Stephenson, Lewis, Lewis and Dibley, Smith and Lewis, Solomon, Dix and Leatherman, Norton and Pasek among others. The activities of insects in shelterbelt plantations have also been subject of study, and there have been a few studies on hedges.

The insect pests of an agroforestry system are essentially the pests of its components: the crops and woody perennials. The dynamics of insect pests and their natural enemies are be governed by the complexity and composition of the agroforestry system. The pest situation in these systems will be influenced by the degree of interaction between the components, the type of agroforestry system and the composition of the plant communities in each component.

Interactions among the components of the agroforestry system can be either positive, negative or neutral. They are regarded as negative when pest problems are increased in an agroforestry system when compared with a monoculture block plantation. A reduction in pest activity under agroforestry indicates a positive interaction, while no change in pest intensities between monoculture and agroforestry denotes a neutral interaction from the insect-pest management point of view. Many factors govern insect-pest intensity in agroforestry and each factor may have a different effect on pests at different times and under different situations. The net outcome will be the sum of favourable and unfavourable effects on pests and their natural enemies.

Vegetational Diversity

Uniformity in plant genetic material has been recognized as one of the main causes of an increase in pest problems in monoculture fields. A large number of experiments carried out under different conditions indicated a reduction in pest activity with diverse vegetation as compared with monoculture.

There are several reviews of insect abundance in diversified vegetation. Studies on the effect of multiple cropping patterns have been carried out by Marcovitch, Dempster and Coaker, Litsinger and Moody, Perrin, van Emden, Altieri et al., Kroh and Beaver, Risch, Altieri and Altieri et al. There have also been some trials with perennial orchard plants.

Weeds in crop fields may affect the activity of insects on the crops. A number of experiments have been carried out to assess this effect.

Various workers have attempted to elucidate the ecological mechanisms underlying differences in the dynamics of insect herbivores and their natural enemies in simple and diverse crop habitats. To explain the general reduction of pest densities in diverse plant combinations, Root proposed two hypotheses, the resource-concentration hypothesis and the enemies hypothesis. The resource-concentration hypothesis suggests that in monoculture fields where the same plant species is cultivated over large areas the herbivores find a concentrated source of food in one place that supports uninterrupted population build up. The food plants in pure stands are easily detected and colonized. The pests, particularly the specialists, exhibit longer tenure periods and higher feeding and reproductive success.

Agroforestry introduces plant diversity in a land unit, over both time and space. Complex agroforestry systems may be close to though not equivalent to natural plant communities in a stable ecosystem or a system in ecological succession. In the latter, the type and pattern of vegetation is governed by the forces of nature. The plant communities developing through natural selection have a degree of in-built resistance to insect attack. In agroforestry, however, the choice of vegetation is determined by people and depends on the objective of the system being practised. This freedom to introduce selective diversity enables agroforesters to choose plants with the desired attributes for accomplishing their objectives.

Different degrees of insect injury occur when a host plant is raised with different companion plants. A reduction in pest numbers and increase in predators was observed when blackgram (Vigna mungo) was intercropped with sorghum or pigeonpea, while intercropping with greengram (Vigna radiata) provided favourable conditions for an increase in pest numbers. The mite populations on cassava were higher in a eucalyptus-cassava combination than in a banana-cassava combination in experiments conducted by Ghosh et al. The grasshopper populations in fields of pearlmillet and sorghum with interspersed neem trees were lower than those in fields with Acacia arabica. Six years after eucalyptus trees were introduced in Malnad, India, a survey revealed that the number of insect species was reduced to a quarter compared with that in areas where no eucalyptus were present. Interplanting beans or allowing weeds to grow with collards considerably decreased flea beetle densities on the collards and minimized leaf damage.

Insect population dynamics are greatly influenced by the type of vegetation in any plant assemblage. Generally, pest levels are not reduced to the same degree in polyculture systems. Polyphagous pests exhibit varying levels of activity on different plants in an assemblage. Studies on the biology of Diacrisis oblique on different host plants demonstrated significant differences in growth and fecundity of the pest. Similar results have been reported in respect of Heliothis armigera. The type of vegetation in a field also affects the activities of the natural enemies of insect pests. In Israel, the cottony cushion scale, leery a purchasi, was found to be resistant to predation by Rodolia cardinals on Spartum junceum and Erythrina corallodendrum plants, whereas the predator is known to be an efficient biocontrol agent of the scale on other plants elsewhere.

Diversity does not always result in reduced pest populations. It appears to be pest specific and also site specific, as well as being affected by other factors. Weed diversity reduced the incidence of fall armyworm, Spodoptera frugiperda, but not of the earworm, Herliothis zea, in a corn field. The intensity of Trachylepida sp attack on Cassia fistula seeds was less in isolated plants as compared with that in mixed stands, indicating that diversity does not always result in a reduction in pest attack.

Many multipurpose woody perennials used in agroforestry possess the inherent properties of wild plants, including genetic diversity, as they have not been domesticated for a long time and have not been subjected to rigorous genetic selection, unlike most other plants of economic importance. Before species considered for agroforestry are subjected to breeding for improvement, their insect-resistance characters should be studied in different provenances.

Taxonomic Alliance

Plants belonging to the same or a very close taxonomic group have the tendency to share common pests. In agroforestry systems, aligophagous and polyphagous insect pests are expected to thrive if both components belong to the same or a closely allied taxonomic group. A large number of plants considered for agroforestry are legumes. Combining crops such as pulses or some oilseeds with leguminous woody perennials may result in supporting pest populations common to both components. The most common example is that of the bruchids. Species of caryedon beetles such as Caryedon serratus infest groundnut as well as a number of legume tree seeds. During the early establishment phase of agroforestry systems, tree components arc likcly to suffer the major injury with attack of the seeds in the limited pods produced. Later, the crops are likely to be more affected as the bruchid populations built up on the tree seeds begin to infest the pulse crops. The damage caused by the pod borer, Etiella zinckelia, in peas has been reported to be accentuated by the presence of acacia trees in the vicinity, as reported by Szeoke and Takacs.

Plant species belonging to the same taxonomic group may contain common or closely related biochemicals that are sought by the insect pests. An insect feeding on a plant with a certain biochemical make-up will adapt more easily to closely related plants with similar biochemical constituents than to species that have entirely different constituents because of taxonomic differences. An agroforestry system comprising plant species belonging to different taxonomic groups is expected to be less affected by insect pests than a system composed of closely related species.

The availability of food over an extended period afforded by closely related plants in the agroforestry field contributes to multiplication of pest populations. The perennial plants in the system may provide a year-round food supply for the pests and thus favour maintenance of the pest population between seasons when the main food—the crop—is absent from the field. It will also result in the maintenance of the population of the pests' natural enemies.

Thus, when considering species for agroforestry, it is advisable to include plants from different taxonomic groups to avoid sharing common pests. The advantages accruing from the maintenance of natural-enemy populations also need to be weighed up.

Non-taxonomic Alliance

Under natural conditions, even insects with a limited host range have been observed to feed on taxonomically diverse species of plants. Thus, taxonomically distant plant species could also be hosts for insect pests. Cocoa grown under leucaena shade suffered more from attacks of defoliating Lepidoptera than when it was grown under some forest trees because the pests were able to use leucaena as an alternative food source. Eucalyptus grown as shade trees with tea shares attack by Chrysolampra flavipes in Assam, India. Cross infestation of cassava mites, Tetranychus spp, on banana and of thrips, Retithrips syriacus, on leucaena have been reported by Ghosh et al. In Hawaii, the thrips, Frankliniella occidentalis, which is a vector of spotted wilt virus on lettuce, tomato, cabbage, etc., occurred on leucaena blossom. The bagworm, Oiketicus kirbyi, a pest of coffee, has been demonstrated to develop on eucalyptus leaves in Brazil. In an agroforestry system, therefore, the plant assembly should consist of species that do not double as host for insect pests of other plants in the system—whether crops or woody perennials.

Some insects utilize different host plants as food in their larval stages from those eaten in the adult stage. Thus an even greater range of plants in an agroforestry system may be attacked by different stages of an insect pest. A plant believed to be a non-host for an insect pest in one stage may turn out to be a host plant for another stage. An example of such a situation occurs in the co-cultivation of crops such as pearlmillet, cowpea, greengram, mothbean and sesame with such tree species as Prosopis cineraria, Azadirachta indica and Zizyphus spp in the rainy season in semi-arid regions of India. The woody species are host plants for adult chafer beetles (Lachnosterna spp, Holotrichia spp, Anomala spp, Adoretus spp and others) whose larvae cause heavy losses to these rainy-season crops in some areas. The presence of such host plants in the crop fields serves to attract the adult beetles. Although the feeding time on the woody plants is very short, the eggs released by the gravid beetles hatch into white grubs, which infest the crops, causing heavy losses.

The insect vectors of plant diseases often survive on a variety of host plants that are not taxonomically close. White flies (aleurodids) and some aphid and jassid species that are vectors of viral, bacterial and other plant diseases often attack a number of different types of plants. The diseases may not affect all these collateral or refuge hosts, but the survival and multiplication of the carrier insect is favoured by them. The presence of refuge host plants in an agroforestry system may provide a source of infection for the crops or human beings if the plant harbours vectors that spread human disease.

The presence of component plants in a system that may serve as host to a vulnerable stage of the insect pest, or that act as a collateral host for plant, animal or human disease, may be utilized to trap and kill these insects while keeping the main crop untreated. Such plants are, however, different from trap plants that are specifically grown to attract insects from the main crop and are not otherwise a component of the agroforestry system.

When establishing an agroforestry system, it is useful to consider whether the component plants serve as host to any vulnerable stage of an important insect pest of the other component plants or as collateral or refuge hosts for the vectors of diseases. Exclusion of such plants from the system could help minimize insect and disease attack on crops or woody perennials.

The Host Range of Pests

The severity of pest infestation in an agroforestry system will depend on the host range of the attacking pests and their relative abundance in the system. Insects with a wide host range will be able to multiply on a number of host plants, while monophagous insects will be restricted to a limited number of host plants within the system. If all or most plants in a mixed system are palatable to a polyphagous pest, then it is likely that the insect will stay longer and become more numerous, causing greater damage. Populations of Empoasca krameri were reduced when beans were interplanted with non-host grasses; but were not affected by the presence of Amaranthus dubius. The abundance of the leafhopper Scaphytopius acutus increased in peach orchards when the ground cover consisted of host plants but was not affected when a non-host grass was used.

If the insect is a relatively specialized feeder, the population density will be lower in plant communities with a higher diversity. On the other hand, the more general pests are likely to increase in abundance if increasing plant diversity in an agroforestry field increases the number of potential host plants in the system. Even for relatively oligophagous pests, increasing diversity may provide a greater number of suitable hosts and lead to increased abundance. A hypothesis to explain lower numbers of herbivores in diverse plant communities proposes that the rate of emigration, rather than the rate of colonization, is the factor most affected by the presence of non-host plants in the system. Studies of various leaf beetles indicate that they have a shorter residence time in plant patches with non-host plants and move farther after encountering a non-host plant. Saxena and Basit found that both colonization and residence time of the leafhopper Amrasca devastans on a host plant were influenced by the presence of the nonhost plants.

The presence of non-host plants in an agroforestry system is of paramount importance in managing pests. In fact, the major variable determining herbivore abundance is the ratio of host to non-host plants rather than the actual number of plant species in an agroforestry system. Through entomological investigations it is possible to manipulate the proportions of host and non-host plants in any agroforestry system. Intelligent manipulation of crop and woodyperennial combinations can minimize insect damage in an agroforestry system.

Monophagous pests can be controlled altogether by not including their host plants in the system. The host range of oligophagous and polyphagous pests can also be narrowed by eliminating palatable species from the assemblage and replacing them with non-host plants. Monophagous pests are most easily managed in this way. In a review of polyculture systems, Andow reported that monophagous herbivores are more likely to decrease in diverse systems than polyphagous pests (61.3% versus 27.1%) and less likely to increase (10% versus 43.8%).

Biological Control Potential

Agroforestry systems, particularly the complex ones, have a great potential for controlling pest populations through increasing the efficiency of biological control agents. The natural-enemy hypothesis proposed by Root to explain reduced herbivory in polyculture systems has been tested in a large number of field experiments and the results were mostly supportive of this hypothesis. Polycultural systems such as agroforestry offer alternate prey, nectar sources and suitable micro-habitats for parasitoids and predators. Being perennial, these systems support the natural enemy population within and between seasons, especially during the off season of the main crop.

Greater colonization and abundance of natural enemies in a mixed culture of plants has been demonstrated in many experiments. Smith studied the colonization of a brussels sprouts field by Anthocorus nemorum with and without weedy vegetation. The predators were more abundant in samples from weedy areas (representing diverse vegetation). Rapid colonization or higher densities of predator species in dense vegetation have been reported by Sprenkel et al. and Horn. Letourneau and Altieri and Letourneau suggested that predator-colonization rates could be manipulated through vegetational diversification of the crop habitat. Similarly, an increase in the diversity of tree species might increase food sources for adult parasitoids.

The presence of alternate prey associated with non-crop vegetation can prevent the local extinction of predator species or increase the proximity of colonizer sources. The population of Plutella maculipennis on cabbage raised with a background vegetation of Crataegus sp was regulated by the parasitic wasp Horogenus sp, for which the background vegetation served as an alternate host. The predating activities of ground beetles were enhanced when cabbage was undersown with white and red clover, resulting in regulation of populations of Erioischia brassicae, Brevicoryne brassicae and Pieris rapae. Regulation of populations of Mamestra brassicae, Evergastis forficalis and Brevicoryne brassicae due to predator enhancement as a result of co-cultivation of brussels sprouts with Spergula arvensis has been reported by Theunissen and den Ouden. Increased abundance

Whether they act as physical or biological barriers, the woody species in an agroforestry system will restrict the movement of insects through the system. This situation is advantageous if the entry of the insect pests is blocked or if outward movement of the natural enemies of the pests is hindered.

Field Configuration and Design

Some insect species, while flying high in the air, recognize their host plants by the field configuration. The configuration of a land unit is determined by the type of plants, their colour, structure, height, density in the field and the type and colour of the soil in the background. Any change in these characters may change the configuration of the field, which will affect its recognition by the insects. Dempster and Coaker observed that even small changes in cropping practices can greatly alter the attractiveness of plots to pests and their natural enemies.

Through the introduction of heterogeneity in a field, it is possible to reduce the number of insect visitors to the field. Heterogeneity can be introduced into the field in several ways, such as through changes in plant density or by mixing different types of plants. The companion crops provide a camouflage for the host crops, thus preventing the pests from recognizing the host from a distance. Some pests prefer plants of a particular colour or texture. In the Philippines, corn borers were found to avoid a green colour on the ground in corn fields. Aphids were found to colonize plants more readily when they stood against a background of bare soil. The attractiveness of collards to aphids decreased when these were grown against a vetch background. Smith found that colonization of Pieris rapae on brussels sprouts was reduced when weeds were allowed to grow with the crop. Similar observations have been reported by Dempster. Dempster and Coaker found that the colonization of cabbages by Erioischia brassicae, Brevicoryne brassicae and Pieris rapae was greatly interfered with when the cabbages were undersown with white and red clover. Colonization of brussels sprouts by Mamestra brassicae, Evergestis forficalis and Brevicoryne brassicae was adversely affected when the crop was grown with Spergula arvensis, probably because of a configurational difference.

Plant-community structures also affect the biology of herbivorous insects and their natural enemies. The number of eggs laid by the pyralid, Cactoblastis cactorum, was affected by plant size, cladode condition, the conspicuousness of the plant, and height above ground. Leigh et al. found plant density to be positively correlated with densities of the predator Orius tristicolor in cotton fields. Carabaeid beetles were more destructive of Pieris rapae caterpillars in weedy plots than in the weeded plots of brussels sprouts.

The arrangement of the different plant species in an agroforestry system may have a profound effect on the activities of insect pests and their enemies. The situation of the crops and woody perennials with respect to the direction of wind and sun determines the effect of these agencies on the dynamics of visiting insects. The effect of shading should be minimal in systems where trees have been planted in alignment with the

of predators in collards with a weedy background checked the growth of Myzus persicae. A greater potential for pest control in complex systems as compared with simple ones was demonstrated by an increased colonization rate by a generalist predator in experiments involving the flower thrip Frankliniella occidentalis and the predator bug Orius tristicolor. The same author reported increased visits of hymenopteran parasitoids in mixed-crop assemblages as compared with pure stands of squash.

Polycultures, especially those containing flowering trees and shrubs, can provide more pollen and nectar sources attractive to and sustaining predators than monocultures. Ageratwn conysoides bears flowers all year round, providing pollen for mites and favouring colonization and build up in citrus orchards. Introduction of flowering perennials or short-lived plants in an agroforestry system will contribute towards biological control of pests.

Some plants produce chemicals that enhance the efficacy of predators. Parasitization of corn earworm eggs by wild Trichogramma sp wasps was promoted by applying extracts of the weed Amaranthus sp. The presence of these plants that increase the predatory activities of insects by virtue of producing such chemicals will benefit an agroforestry system. The parasitization of Cotesis kazak was more serious on Heliothis armigera larvae feeding on cotton, okra and tomato, than on those feeding on chickpea, pigeonpea, cowpea and dolichos bean under similar circumstances.

A major factor determining natural-enemy abundance in mixed vegetation is the suitability of the microhabitat provided by one or more of the plants in the assemblage. Trees and shrubs often provide better shelter and mating sites than do short-lived annual plants. Hedges provide very favourable environments for parasitic Hymenoptera and Diptera. The Braconidae tend to colonize the leeward side of hedges and other thick vegetation, while the Vespidae and some Diptera accumulate more on the windward side. These plants also provide pupation, mating and over-wintering sites for natural enemies. Many predatory spiders prefer to inhabitate woody plants rather than annual plants. The web density of the spider Stegduphus sarasinerum was higher on such plants as Acacia tortilis, Prosopis cineraria, Capparis sp, Tecomella spp and Zizyphus spp than on less woody species.

The presence of different herbivores in an agroforestry system may encourage predators to remain when their main prey is rare. Prey densities that fall below a certain threshold may cause emigration of natural enemies from an area. It is, therefore, important that prey availability is maintained in the system. Predators often have a wider host range than parasites and thus have a better chance of survival in the event of the population of the main prey falling to a low level.

Microclimate

The interactions among plants in a mixed agroforestry system create a microenvironment that is different from that of the surrounding area. Shade is the most prominent consequence of tree-crop combinations and it may have both a direct and an indirect effect on the activity of pests and natural enemies. Among other effects, shade provides protection from direct sunlight and regulation of light intensity underneath the tree. Reduction in temperature and an increase in humidity are indirect effects. Most diurnal insects, especially soft-bodied larval stages, prefer to feed while avoiding direct exposure to the sun. Under shady conditions their feeding activity could be enhanced, thus increasing their damage potential. Light intensity governs the habitation of fields by some dung beetles. Most aphid species prefer shady conditions in the warm climates. Some pests prefer sunny conditions. Risch observed that the density of beetles on beans under the shade of maize plants was less than on beans in monoculture. Infestation by Polyura naraea was higher on stands at the edges of forest canopy facing the sun than on the shady side. Shade from trees can interfere with the hostseeking and reproductive behaviour of some insects. Many hymenopteran parasites exhibit greater host-searching capacity under bright-light conditions, so their activity may be retarded in the shade. Whether shade has a positive or a negative effect on pest activity depends on the habits of the pests in the agroforestry system. The shade intensity will depend on the growth habits, structure and arrangement of the trees in the system.

The humid conditions in an agroforestry system may be favourable for the development of disease in insect pests. Coupled with the absence of direct sun, the effectiveness of entomopathogenic fungi may be increased by humidity. In experiments on the effect of sunlight on the field persistence of Nomuraea rileyi, Fargues et al. found that the half-life and viability of spores of this fungus were much longer in the shade. The effective infective period of Bacillus thuringiensis, the most widely used entomopathogenic bacterial preparation, is greatly reduced in sunlight. Increased humidity has also been reported to favour parasitization of pest eggs by Trichogramma sp.

Masking Effect

Many plants emit volatile chemicals or odours into the environment. These odours are perceived by herbivorous insects and utilized to orient themselves to the host fields. Onion flies, Hylemya antigua, have been found to fly directly towards the odour released by an onion bait. In monocultures, the odours released by the plants spread out in all directions and are perceived by the herbivores without any interruption. But in mixed vegetation, the odours released by some plants may mask the effect of those released by other plants. Under these circumstances, the insects find it difficult to locate the hosts on which to feed and reproduce.

Thus, odoriferous plants, when raised with host plants of insect pests, can deter recognition, feeding and reproduction of the pests on their host plants. Tahvanainen and Root attributed the lower density of Phyllotreta cruciferae on cabbage raised with tomato, tobacco and ragweed to feeding inhibition by odours from the nonhost plants. The lower activity of Plutella xylostella on cabbage intercropped with tomato has been attributed to the repellent effect of the chemical produced by the companion crops.

Many plant residues contain allelopathic compounds that could affect the growt adjoining plants in a mixture, changing the attractiveness of the plants to their p In experiments with living mulches, Altieri et al. observed lower aphid coloniza and population build-up on collards grown with vetch (which they suspected to con allelopathic compounds) as compared with that on collards in monoculture plots.

The masking effect is not restricted to the insect pests. The performance of n ral enemies may also be affected by chemical cues from the associated plants. prey-seeking behaviour of insect pests that utilize the odours of the host plant their prey to locate them is influenced by the presence of other vegetation in system. Such a situation will reduce the biological-control potential of the sys However, there could also be cues from companion plants that help establish nat enemies in the system. Identification of the plants that release pest-antagonistic natural-enemyattractive odours may be of great importance in developing a pest- agroforestry system.

Barrier Effects

The tall woody plants in an agroforestry system act as a physical barrier to the m ment of insects to, from and within the system. The non-host plants of insect raised with their host plants act as biological barriers, restricting their movemen wards the host plants. Hedges, boundary plantations and windbreaks affect the nization and dispersal of both herbivorous and predatory parasitic insects. Both zontal and vertical movements of insects are affected by tall or thick woody plantat The permeability of the vegetation affects the movement of insects in and out o system. Inand-out migration patterns of insects in polycultures have been studie Bach, Risch, and Wetzler and Risch.

Non-host plants mixed in with host plants either act as a mechanical barrier t dispersal of the insect pest or physically repel the pests because of unpleasant phological features such as hairy leaves. Weakly flying insects such as aphids, th flies and small beetles are carried far afield by the wind. Once in the air, these in cannot land directly on their host-plant plots because of the high speed of the carrying them. They utilize tall trees as obstructions to settle on and later move t crop fields. Thus, the trees in an agroforestry system may serve as agents for facilit colonization of insect pests by providing a platform from which they initiate flig infest the crops.

The woody plants in hedges and boundaries serve as reservoirs for insects. The abundant taxa within hedges are parasitic Hymenoptera and Diptera. Although he contribute to maintaining some populations of insects, most of the increase in i density results from insects blown in from elsewhere. The Braconidae tend to accu late on the leeward side, while the Aphididae, Vespidae, large Diptera and Lepidop predominate on the windward side.

sun's path. Crops raised on the windward side of boundary plantations are less likely to be infested by insect species that are carried by wind.

The design of an agroforestry system will define the extent of the influence of weather on insects. The effect will be the least in systems that have thick vegetation vertically and horizontally. The pest situation in agroforestry may be manipulated through suitable design of systems so that the conditions are favourable for the predatory insects and unfavourable for the pests. The size and density of plants and their relative arrangement in time and space will determine the sphere of activity of the insects.

Exotic Plants and Pests

Plants with desirable attributes are often moved from one place to another. In the absence of proper phytosanitary measures and non-compliance with quarantine rules, these plants carry their pests with them to areas where they did not occur previously. In the absence of natural enemies at the new location, these insect species rapidly multiply and establish themselves as serious pests in the new location. The psyllid Ctenartaina eucalypti, the curculionid Gonipterus scutellatus and the cerambycid Phorocantha semipunctata have moved with eucalyptus and established themselves where the tree was introduced. The margarodid Auloicerya acaciae was accidentally introduced into Australia, where it became established on a number of acacia species. The Australian sawfly, Phylecteophaga froggati, is causing damage to oak and eucalyptus in New Zealand. Of the 500 species recorded on agricultural crops and forestry plantations in the former USSR, aproximately 80 were introduced insects. In Africa, Rastrococcus invadens is becoming established on horticultural and forest plants after its introduction with plant material.

When an exotic plant is introduced into a locality, it may affect or be affected by the local insect fauna in a number of ways:

- The introduced plant may become a host for established pests in the area. After it was introduced into Australia, Delonix regia became a host to the longicorn Aridaeus thoracicus, a local pest of pear. Acacia tortilis, an introduced plant in India, is attacked by a local pest, Julodis sp, in desert areas. A number of local insects have adapted to eucalyptus in areas where it has been introduced all over the world. In Hawaii, the thrips Frankliniella occidentalis became established on leucaena after the introduction of the latter on the island.
- The introduced plant may become a suitable host for insect species that were not considered important pests on local plants. When such insects multiply and establish on the exotic plant, they acquire the status of pests. Before the introduction of eucalyptus into India, Celosterna scabrator was considered a minor pest affecting the acacias. It adapted to eucalyptus and is now considered an important pest on that genus.

- Exotic plant materials may bring with them some of the pest insects from their native habitat. Or the insects may arrive accidentally in the area where the exotic plant is introduced. In the absence of natural enemies at the new locations, these insects may multiply very fast and become established as major pests. Leucaena leucocephala, introduced in several parts of the world, is heavily infested by the psyllid Heteropsylla cubana, which arrived accidentally at the new sites. The cypress aphid, Cinara cupressi, is causing extensive damage to cypress in Africa.
- An exotic pest coming in with exotic plant material or being accidentally introduced later may establish on other plants in the location in addition to the original host plant. The leucaena psyllid, Heteropsylla cubana, has been found feeding on Samania soman in Asia.

When an exotic plant is introduced in a locality, there may be an interval before it begins to be attacked by insects. It is necessary to observe the plant for at least two years before the insect fauna establishing on it can be determined.

Domestication of Plants

A number of woody plants used in agroforestry have not been improved. In their natural habitats they withstand insect attack through in-built characteristics such as predator satiation and natural resistance. When these plants are brought into agroforestry/ the naturally occurring insects can now be considered pests as each plant acquires economic significance.

Introduced non-domesticated woody plants can sometimes become pests— that is, weeds. The woody plants that flower and fruit early and profusely may disseminate their seeds in the crops and interfere with their growth. Leucaena plants have this tendency and have already become weeds in some highrainfall areas. Such plants are best pruned before flowering or fruiting to avoid dissemination of the seeds to cropped areas. The insect species that are pests on these woody plants turned weeds may be used as biocontrol agents in the affected areas. Many acacias introduced to Australia and South Africa have become weeds, and farmers are seeking to control them using insects. Some species of sesbania are also considered weeds in South Africa. Insects acquire additional significance under such conditions.

Tree-crop Competition and Nutrition

The component plants in a mixed system vie for essential resources. Although agroforestry is envisaged as a system of plant species that benefit each other mutually or unilaterally, it is too optimistic to assume that all types of competition can be eliminated in these systems, especially in areas with poor soils and scanty rainfall. Agroforestry systems lose some of the assimilated nutrients in the form of grain, wood, fodder, etc., at each harvest, thus reducing reserves, unlike natural forests where recycling of nutrients occurs. Insects play an important role in removing plant materials from agroforestry systems. In studies with Leucaena leucocephala, Gliricidia sepium and Flemingia sp leaf mulches, Budelman found that soil-dwelling insects and other arthropods were responsible for considerable losses of mulch. Reviewing various hypotheses concerning the advantages of insect herbivores for plants, Lamb concluded that high levels of herbivory are unlikely to be nutritionally beneficial to the grazed plants.

The general condition of a plant affects its susceptibility to insect attack. Termites preferentially attack water-stressed plants or those stressed in other ways, although they sometimes also attack healthy plants. The same level of infestation may bring about different degrees of injury in the healthy and stressed plants. Tuset and Hinarejos observed that water stress increases susceptibility to bark invasion by Pestalotiopsis junerea in cypress and palms. Leaf toughness in eucalyptus influenced population build-up of the chrysomelid beetle Paropsis atomaria.

Nutrient concentration affects the attractiveness of plants to invaders. Ohmart et al. found that nitrogen plays an important role in the population dynamics of chrysomelid beetles on eucalyptus. In agroforestry, nitrogen-fixing trees make available more nitrogen to the crop components, which may enhance their attractiveness to insect pests. Mycorrhizal inoculation may partly take care of phosphorus utilization, but other nutrients and micronutrients may have to be supplemented to counter the effects of excessive nitrogen levels from a pest-control point of view.

Management Practices

The agronomic practices involved in agroforestry system management may affect the activities of pests and their natural enemies. An undisturbed system favours the establishment of a stable pest-natural enemy equilibrium. In intensely managed systems the parasites and predators are more likely to be disturbed than the pests. Emigration of the natural enemies as a result of disturbances may lessen the benefits of biological control in the system.

Mulches in an agroforestry field may serve as a source of food for many soil-dwelling insects. Budelman observed that a large part of the mulch was destroyed by soil-dwelling arthropods before it decomposed.

Severe pruning of plants in alley cropping gives rise to new vegetative growth, which is more palatable to the insects than the older growth. Removal of infested plant parts by pruning may help to get rid of sessile sucking pests. Fallowing part or the whole of a unit of land affects insect abundance in the unit, especially of the soil-dwelling stages. Intensification of fallow resulted in increased grasshopper numbers, while cultivation and afforestation affected the grasshoppers adversely in investigations carried out by Amatobi et al.

Strategies for Pest Management in Agroforestry

There are inherent advantages in agroforestry from an insect-pest management point of view. In order to get the most out of these practices, the features of an agroforestry system that influence insect pests should be suitably manipulated. A properly designed agroforestry system with the right choice of components can create conditions that are favourable for natural enemies of the insect pests and unfavourable for the latter. The establishment of an environment-friendly agroforestry system requires that the following factors be taken into account.

Choice of Species

To introduce selective diversity in an agroforestry system, the plants chosen should be taxonomically far apart, have a narrow pest complex and have resistant strains. The mix of species in the assemblage should be altered until the right combination of plants is obtained as indicated by a reduced number of pests and an increased number of predators. Such a system would be stable from the insect-pest point of view.

Microclimate

When introduced into a land unit, trees create a microclimate in the area that is different from that of the surrounding areas. Depending on the tree species chosen (whether erect, spreading, with a thick or thin canopy cover, etc.) and the crops underneath, a microclimate of the desired type can be created for the agroforestry system. The choice of tree species is determined by the requirements of the crops concerned. Systems with multistrata vegetation are expected to favour the development of food webs, thus creating a stable pest-parasite equilibrium.

Field Configuration and Design

The arrangement of crops and woody perennials in an agroforestry system affects the activity of insect pests and their natural enemies. By planting trees in line with the direction of the sun's path, the shading effect can be minimized, while planting perpendicularly to the sun's rays will give maximum shading. Where the trees are planted will depend on whether the crops require shade or sunlight. Plants aligned in the direction of the wind will let insects pass through the field, while those aligned across the wind direction will shield the field.

Introduction of Barriers

Large trees act as barriers to the movement of wind and insects in a field. The non-host plants may also act as a barrier to pests. These may be planted in broad strips along with the host plants. Thick hedges around the field define the boundaries for the movement of natural enemies of the pests. Preferably hedges should be of flower-bearing plants to provide food for the natural enemies.

Odoriferous Plants

Introduction of a variety of odour-emitting plants in an agroforestry system may help to divert the major pests if they are guided by odours. The system can be planned to include plants that would repel the pests of companion crops.

Trap Plants

In the case of some pests, the main crop or woody plant may be protected from attack by insect pests by growing a more palatable plant with it. Such 'trap' plants may or may not be component plants of the system but serve to divert the pests from settling on the main crops.

Management Practices

The practices for management of an agroforestry system need to be developed or modified with a view to minimizing insect-pest incidence and injury. Depending on the habits of the major insect pests in the system, agronomic practices may be tailored to discourage the pests and to favour the predators and parasites.

There are several ways in which pest populations may be kept down in an agroforestry system, though it is difficult to test all of them in the field. The practitioner can develop an effective package after trying different practices.

WATERSHED MANAGEMENT IN AGROFORESTRY

Watershed management as the process of guiding and organizing land use and the use of other resources in a watershed in order sustainably to provide desired goods and services to the people without adversely affecting soil and water resources. Watershed management is generally scientifically management of water in the earth.it has evolved and passed through several developmental stages. In the initial stages, it was a subject of forestry and forestryrelated hydrology. The involvement of people was not an issue. It was solely an affair of government forest departments. During the second stage, it became land resources management-related, including activities with an eye on economic benefits. At this stage, the focus was on beneficiaries. It is now –participatory and integrated watershed management, with involvement and contribution from local people.

Worldwide environmental, socio-economic and political changes are challenging some of the foundations on which watershed management has been based for the last 25 years. Watershed management is going through a period of experimentation in which old and new practices often coexist and mix.

The new generation of watershed management programmes being developed has a different approach, design and implementation strategy.

New Integrated Watershed Management Model for Efficient Management of Natural Resources

A new model for efficient management of natural resources in the SAT has emerged from the lessons learnt from long watershed-based research. The important components of the new integrated watershed management model are:

- Farmer participatory approach through cooperation model and not through contractual model.
- Continuous monitoring and evaluation by the stakeholders.
- Emphasize on individual farmer-based conservation measures for increasing productivity of individual farms along with community-based soil and water conservation measures.
- A consortium of institutions for technical backstopping of the on-farm watersheds.
- A holistic system's approach to improve livelihoods of people and not merely conservation of soil and water.
- Link on-station and on-farm watersheds.
- Use of new science tools for management and monitoring of watersheds. Empowerment of community individuals and strengthening of village institutions for managing natural watersheds.
- Low-cost soil and water conservation measures and structures.
- Amalgamation of traditional knowledge and new knowledge for efficient management of natural resources.
- A micro-watershed within the watershed where farmers conduct strategic research with technical guidance from the scientists. Minimize free supply of inputs for undertaking evaluation of technologies.

Role of Agroforestry in Watershed Management

Forests are the biggest users of water worldwide and extensive forested areas have been lost or are undergoing conversion to agriculture, creating concerns about loss of hydrological functions and increasing the competition for scarce water between agriculture, urban centers, industries and wildlife. The challenge is to improve the sustainability and productivity of land and water use, especially for the growing populations of many developing countries. Recent findings on the hydrology of forests and agroforestry systems and indicate how modifications in tree based systems might increase water productivity. In forestry, the focus of research has moved from the hydrological functions of upland forest reserves that are close to settlements to a greater recognition of the roles played by upland communities in the management of water resources. A major source of conflict over water resources is the contrasting perceptions of watershed functions' between forest managers and local people, which are often based more on myths of forest functions than on science – for example, the idea that forests increase rainfall. These myths continue to dominate the views of policy makers and institutions and should be revised. The challenge is to gain a better insight into how farmer-developed landuse mosaics have modified watershed-protection functions. Priority must be given to the perceptions, experiences and strategies of local communities.

Trees on farms have the potential for improving productivity in two ways. Trees can increase the amount of water that is used on farm as tree or crop transpiration. Trees can also increase the productivity of the water that is used by increasing biomass of trees or crops produced per unit of water used. Plot-level evidence shows that improvements in water productivity as a consequence of modifications to the microclimate of the crop are likely to be limited. Instead, evidence from semi-arid India and Kenya showed that the greater productivity of agroforestry systems is primarily due to the higher amount of water used. Almost half of the total water use occurred during the dry season, when cropping was impossible, and the rest was extracted from soil reserves. This implies a high temporal complementarity between the crop and tree components of the landscape mosaic. Research is needed to examine the impact of the increased water use on the drainage and base flow at the landscape level. This chapter also describes some of the technical approaches that can be used to improve land and water management, the role of trees and its relation to hydrology and the challenges for rational land-use decision-making.

Agroforestry offers one promising option for efficient and sustainable use of land and water. In simplified terms, agroforestry means combining the management of trees with productive agricultural activities. Agroforestry provides opportunities for forest conversion in the true sense of the term – that is, replacement of natural forests with other tree based land-use systems. There are also opportunities to use agroforestry for the prevention or reversal of land degradation in the humid tropics.

There are numerous potential benefits that agroforestry systems can achieve, ranging from diversification of production to improved natural-resources utilization. The key benefits in terms of natural-resources use are as follows:

S.No.	Benefits in Terms of Natural-resources
1.	Soil conservation in terms of protection against erosion.
2.	Improvement or maintenance of soil fertility.

3.	Water conservation and more productive use of water.
4.	Providing environmental functions required for sustainability.

Role of Watershed Management for Climate Change

The world's watersheds face a wide variety of stresses that threaten to degrade their biological value and their ability to provide ecosystem services. Prominent among the pressures faced by watershed include high population growth, expansion of irrigated cropland, high deforestation and acute water scarcity. The stresses are particularly severe in watersheds that are already substantially modified or degraded by human activity in India, China, and Southeast Asia. Other major watersheds that are less degraded, such as the Amazon and the Congo, are nonetheless beginning to experience rapid transformations that threatens livelihood. The accelerating pace of climate change is an additional challenge to the sustainability of global watersheds. Climate change has the potential to damage irreversibly the natural resource base, on which agriculture depends, in many watersheds, which aggravate the consequences for food security. There is a consensus among climate change scientists that major watersheds in Africa and other tropical regions will be impacted to varying degrees by changes to the frequency, intensity and timing of rainfall within seasons, shifts in seasons with wet seasons becoming shorter, dry seasons longer and more uncertain timing in the transitions between them and warmer temperatures increasing evapo-transpiration and impacting on soil water balance. These climate fluctuations necessitate increased attention to addressing human and climatic stresses on watersheds in order to conserve their ecological and economic functions.

Working at watershed or landscape level allows to anticipate the impacts of climate change from upstream to downstream. Indeed climate change is expected to intensify extreme weather events such as floods and droughts whose impacts especially depends on land use. This is expected to make communities both from upstream and downstream even more vulnerable. Because of poverty, many local communities live on degraded watersheds placing them at the center of a vicious circle: degraded ecosystems are more sensitive to erosion and further degradation, which are enforced by frequent floods and droughts. Topsoil erosion degrades water quality downstream, decreases agricultural productivity and further impoverishes the rural communities.

Interest in and awareness of the multiple environmental, economic and social benefits provided by watersheds has greatly increased in recent decades. This is particularly true in developing countries where the economy depends predominantly on agriculture. Besides, most of the developing countries are experiencing degradation of land and water resources at an alarming rate, whereas the need for these resources is vastly increasing. Sustainable use and management of land resources can only be achieved by adopting a system of improved land and water management based on an integrated approach to land resources development, and with the direct involvement and participation of the different actors.

Carbon sequestration refers to the capacity of agricultural lands and forests to remove carbon dioxide from the atmosphere in a manner that is not immediately remitted into the atmosphere. Carbon dioxide is absorbed by vegetation through photosynthesis and stored as carbon in biomass and soils. Forests and grasslands are carbon sinks because they can store large amounts of carbon in their vegetation and root systems for a relatively long period of time. However, soils are the largest terrestrial sink for carbon and the ability of agriculture lands to sequester carbon depends on climate, soil type, vegetation cover and land management practices Soil carbon sequestration allows for the replenishment of soil organic matter and, provides several other benefits including improved soil structure and stability that leads to reduced soil erosion, improved soil biodiversity, increased nutrient holding capacity, increased nutrient use efficiency, increased water holding capacity, increased crop yields and profitability and potential marketability of the sequestered carbon. Soil carbon sequestration is good for the soil quality, both at short-term and long-term. It is a cost-effective and environmentally-friendly process that can be achieved through land management practices adapted to specific land uses. Once sequestered, carbon generally remains in the soil as long as the sustainable land management practices are followed.

Carbon Sequestration and Watershed Management

Carbon sequestration at the heart of Climate Smart agricultural policies. Global surface temperatures have increased by 0.8 °C since the late 19th Century with an average rate of increase of 0.15 °C per decade since 1975. The Earth‘s mean temperature is projected to increase by 1.5 °C – 5.8 °C during the 21st Century. Future global warming will exacerbate hydrologic scarcity and variability such that crops will have to grow in warmer and drier conditions. Higher temperatures and shorter growing seasons will reduce the yields of most food crops, and promote the spread of weeds and pests. Changes in precipitation patterns will also increase the likelihood of short-run crop failures and long-run productivity decline. Although there will be productivity gains in some crops in certain regions of the world, the overall impact of climate change on agriculture is expected to be negative, threatening global food security.

Watershed Management a Tool to Address Climate Change

Watershed management is management of land, water and other natural resources on a sustainable basis. Watershed management carries out number of activities with an integrated approach addressing proper land use, protecting lands from all forms of degradation, building and maintaining soil fertility, conserving water for farm use, proper management of water for drainage, erosion control, production, flood protection, sediment reduction and increasing productivity from all land uses. To fulfill the watershed objectives, different Agronomic/Management, Vegetative, and Small Structural Conservation Techniques are packaged into different activities/measures. These measures may be broadly classified as Preventive, Rehabilitative, Conservation education and extension, and Income generating. Following are some rationales of watershed management activities generally implemented to address the water scarcity problems exaggerated by the climate change.

- Vegetative Measures: Vegetative measures such as degraded land rehabilitation mainly through planting grass, tree and fruit sampling, hedgerows, agro-forestry mainly in degraded barren area is one major activity of watershed management. Vegetated watersheds have a sponge effect. Forest soil, roots and litters act as a giant sponge, soaking up water from rain and runoff and releasing gradually over an extended period.

- Protection of Water Source from Erosion: Land use beyond its capability and without proper conservation measures enhances erosion in uplands and sedimentation downstream. One of the principles of watershed management is to use the land as per its capability or suitability. So that erosion therefore water pollution from sediment remains low. This helps to protect the water source such as springs, lakes/ ponds, and rivers and enhance their suitability for human use.

- Efficient use of Water: Proper use of available water resource is one of the key strategies to address the world water scarcity. Watershed management considers different efficient technologies such as drip irrigation, water efficient conservation farming system such as system of rice intensification, etc.

- Drip Irrigation: Drip irrigation is a method of watering plants by delivering drops of water in a controlled way to plant root zones of the crop. The system is made up of a water tank and a network of pipes with drippers at predetermined intervals.

- Water Harvesting: Water harvesting is a collection and storage of water during excess period and use when required to address the water scarcity problems. Under water harvesting, conservation pond and roof water harvest are most demanded activities.

- Conservation Pond: Conservation pond is an indigenous practice to store wastewater and / or run-off water during excess rain to reduce erosion as well as irrigate the crop during critically stressed period. The primary concern of a conservation pond is to manage water wisely for soil conservation and watershed management purposes. However, its use can be extended to serve multiple purposes: (buffalo bathing), irrigation, fisheries, entertainment and so on as required. Conservation pond (non-lined) plays crucial role in recharging ground water improving soil moisture through regulating infiltration, which increases the production. Depression of the conservation pond will have better moisture condition and can be used for growing crops during the dry period.

Role of Watershed Management for Social Aspects

One of the improvements introduced in recent years is the global approach. Although evaluations of the constraints encountered in and the potential of rural areas revealed the complexity of the situations, they also highlighted the need to take account of the many aspects of the problems to be dealt with. A systemic analysis showed that all human activity may be considered as an element in a complex system; a system where various elements are interrelated and can be exchanged (goods and information) within and outwith the system in accordance with regulatory mechanisms, thereby providing a result while at the same time ensuring that the overall system is sustainable. Under these holistic arrangements, the human factor has a vital role in rural systems. If we look at mountain dwellers, we find that they develop survival strategies based on the assets available to them (land, water, inputs, labour and expertise) and are able to obtain results that can be used not only to sustain their system but also, in the best-case scenario, to expand it. Of course, farmers receive funds in the form of aid and subsidies, as well as non-farm income from family members. Understanding the strategies used by small farmers, as well as how small farming systems work or why they fail, could shed light on the constraints they face and the measures to be taken to overcome them. Such measures could include improving the factors of production (e.g. fertility improvement, improved water management, land management, equipment and mechanization, etc.) and the marketing of farm surpluses (labelling, marketing organization, etc.). This requires decisions to be made by the persons in charge, their families and other persons concerned directly or indirectly with these improvements. All these partners are also involved in a number of rural systems all governed by the same rules of operation (e.g. marketing systems, the banking system, the political system, etc.). While each system may be an entity in itself, all the systems are inter-related and may also be interdependent.

The hydrological system is intrinsic to the watershed, a truly complex system. The same may be said of social systems, which may be local, regional, national and even international. They all involve a number of flows and relationships among the various components which, despite the complexity of the task, must be taken into account so that an accurate evaluation of the environmental situation can be made and the most appropriate solutions recommended.

Similarly, the changes in the design and implementation of development programmes in recent years have made it necessary to review the principles and relationships among the players in the rural process, especially the partnership between the people and the outside experts. Indeed, evaluations of the measures taken, often on the initiative of the authorities and outside experts, have highlighted problems concerning the maintenance and sustainability of the measures taken after project completion. The situation is further aggravated by pressure from international donors for government disengagement. The new responsibility-sharing system is part and parcel of a general trend towards decentralization, now taking place in most countries. It also draws on a very strong movement promoting the involvement of grassroots communities in the sustainable management of their resources.

Participation and Local Development

Although people's participation is recognized as a necessity and has been introduced in many programmes and projects as well as in most national and international plans, it is not always evident that it is being implemented. Some of the problems with involving the people in projects lie with the outside experts who have difficulty changing their method of operation – a management-based and top-down approach – and do not fully understand the reality of the situation in which the people find themselves. The people, on the other hand, find it difficult to enter into a new type of relationship as they continue to see themselves as the recipients and the outside experts as the providers of material assistance. However, what slows things down most often is failure to recognize the local people and their associations as true partners.

Social Relations and Partnership

A policy based on local player involvement needs to recognize the demands of local communities and small regions. Conversely, national policies will have to be made more regionfriendly. In other words, they will have to take into account the regions' agro-ecological, social and cultural characteristics. However, these two dimensions can only be implemented if accompanied by strong support measures designed to improve information sharing, strengthen the capabilities of people at all levels and organize the countryside. The major challenge here is how to deal with the contradiction between acknowledging local community initiatives and the need to incorporate these initiatives into a comprehensive approach.

With decentralization, the State becomes the mobilizer and facilitator of local development initiatives proposed by the communities. The basic idea behind economic reforms and decentralization is to give free rein to initiatives so that they can cater to special interests, without the local elites once again using their role as representatives of the people to organize, run and take over decentralization. A contractual and partnership approach would seek to establish new relations among the rural development players rather than vertical relations based on strategies that ignore local and regional processes. Centralized government institutions must be replaced by new institutions capable of creating suitable conditions for dialogue between farmers' organizations and other rural development players. At the same time, these new institutions must work towards the creation, conversion and strengthening of intermediate associations, which will have a central role for the following three reasons: 1) they will provide guidance for the government in drawing up the various policies that must go hand in hand with decentralization; 2) they will collate and regionalize the demands and requests of the rural people; and 3) as they eventually become sufficiently mature consultative partners, they will be able to build partnerships with other rural development players.

Watershed Development and its Impact on Livelihoods

- Reduction in the threat of drought to crop and livestock production.
- Recharge of ground water.
- Improved fodder production.
- More livestock managed under stall-fed conditions.
- Diversification of the village economy into artisanal and other activities as people gain the confidence to approach banks for credit.
- Increase in cropping intensity and yields of both irrigated and dry land crops.
- Increase in milk production (livestock systems move from open grazing and towards crossbreeds.
- Decline in sedimentation downstream.
- Creation of employment opportunities for landless labour.
- Year-round availability of drinking water.

BAMBOO BASED AGROFORESTRY

Bamboos based agroforestry can play an important role in enhancing productivity, sustainability and resource conservation. Many of the useful bamboo species can occupy the same ecological niche as trees and are well suited for agroforestry. Bamboos have many advantages over trees such as, relatively short time span from planting to harvest versatility of use which outmatches most tree species the ability to provide building materials and edible products for many years or even decades. Its growth rate is three times that of eucalypts and mature in just three years hence, there are large scale efforts to promote bamboos under agroforestry system.

Bamboos require four to five years to yield first harvest, if grown from offsets, which is much earlier than any other woody species. If raised from seedlings, first harvest is obtained after seven years. This initial period can be sustainably utilized for raising intercrops and enhancing sustainability and income of the growers. Under agroforestry system, bamboos are also benefited due to sharing of resources like irrigation, fertilizers, weeding, etc. with intercrops, as a result the quantity and quality of bamboos are expected to be much higher as compared with monoculture and unmanaged plantations. The bamboos based agroforestry practices which indicate that the safest choice of agroforestry species have come from the native vegetation, having a history of adoption to local precipitation regimes. Balaji reported thatthe scope of bamboo in agroforestry is very wide because of the uncertain weather conditions and increasing cost of labour.

Bamboos, if properly managed, can be grown in agrisilviculture, silvipastoral, agrisilvipastoral and agrisilvihorticultural system.

Agroforestry plantations can be raised for socio-economic and ecological considerations.They can be suitably grown forintercropping for enhancing productivity and conservation of natural resources using monoculture plantations, wind breaks, riparian filter, permaculture, etc. Bamboos of different heights and growth characters may be used for integrating them with agriculture crops. The species to be selected for agroforestry models should have light crowns such as Dendrocalamus, Phyllostachys andThyrsostachysspecies. It has been emphasized that basic principlesfor model establishment should have clear management objective, suitable management strategy, positive relationship between the model and environment, and maintain appropriate control of the compatibility and competition relationship among the species and full use of resource.

Bamboo based Agroforestry for Productivity Enhancement

On the basis of experiences from South China, more or less similar to the environment and bamboo types growing in South Asian countries, Fu Maoyi and Banik described bamboo based agroforestry models. Some of the models being practiced in India and worldwide have been categorized and discussed as under:

Bamboos in Homegardens

In homesteads, they are either found mixed with a large number of other tree species or pure in patches. Bamboos in the village homesteads mostly occupy the backyard and periphery of the holdings and are one of the most important crop to the farmers. Most of the species in homegardens are usually clump forming, congested in nature with large, tall branchy culms. Bamboos easily grow as inter- and under- crop with many trees.

Inmoist humid zones of north-east India, West Bengal and Odisha, genus Bambusa is most commonly cultivated. B. balcooa, B. bambos, B. nutans, B. tulda and B. vulgaris are common bamboos species in the homestead in this part. Crops like siris, aonla, bakain, banana, beetelnut, coconut, neem and semul are grown in combination with bamboos. It has been estimated that 15-30 per cent oftotal plant crown presentin homestead in Assam, Tripura and West Bengal contains bamboos. Gangopadhyay surveyed the home gardens in 13 sample villages in five districts of M.P. in India and reported that about 15.13 percent offamilies were having D. strictus on theirfields. In less moist to semi-drier parts of Bihar, Uttar Pradesh, Madhya Pradesh and Maharashtra, B. bambos and D. strictus are commonly cultivated a long with B. balcooa, B. tulda and B. nutans. In cooler parts of north-east region like Tripura, Meghalaya, Tamenglong (Manipur), some areas of Assam, Arunanchal and north Bengal, D. hamiltonii is grown naturally in the moist sites. In Himachal Pradesh and Uttarakhand, D. hamiltonii and B. nutans are grown by farmers in mid hills. Species like bhimal (Grewia optiva), khirak (Celtis australis), shetoot (Morus alba), kachnar (Bauhinia variegata) are also found growing with these bamboo species. In tarai region of Uttarakahnad, Bengali migrants are cultivating B. balcooa, B. nutans, B. tulda and D. hamiltonii with bael, citrus, kathal, moringa, neem and semul in their homesteads. At higher altitudes of Himachal Pradesh and Uttarakhand, temperate species like Drepanostachyum falcatum (ghad ringal), Himalayacalmus falconeri (deo ringal), Thamanocalamus spathiflorus (thaam ringal), Thamnocalamus jaunsarensis (jamura ringal) are usually found. In drier parts like Punjab, western Maharastra, drought tolerant species like B. bambos and D. strictus are cultivated. In Karnatka, Goa, Andhra Pradesh and western Kerala, Pseudoxytenanthera stocksii is cultivated which is not found in other parts of India.

Block Plantation/Agrisilviculture System

In this model, bamboos are planted at a spacing varying from 4 m x 4 m (for small sized bamboos) to 9 m × 9 m (for large sized species). The interspace is utilized for raising annual crops. It has been well documented that bamboos and tree species gradually become more competitive with age and, thus, progressively the crop yield is reduced. Intercropping, with 2 crops employed in bamboos + trees model, should have a bamboo plant to have 1 m in area to ensure adequate moisture and nutrition supply. Intercropping with bamboos, however, can be done for a maximum of four years after bamboo planting, after which there is huge competition for the resources. Bamboo species can be chosen depending on climatic conditions and soil type of the area. In planting bamboos, full soil preparation may be employed on plain land. On sloping land, strip preparation leaving alternate unprepared strips to prevent water and soil erosion is recommended. It is necessary to place adequate fertilizer in the pits before planting. Under agrisilviculture system, soya bean, niger, mustard, wheat, urd and arhar and some of the important intercrops of the initial stages. Raising of shade tolerant crops such as pineapple, ginger, turmeric, shade tolerant variety of sweet potato, cinnamon, etc. within a stand of adult bamboo clumps is technically feasible and economically viable. By adopting wider spacing of bamboos and judicious manipulation of bamboos canopy, the period of intercropping could be extended further. Intercrops can also be taken by keeping large spacing between lines and less spacing between plants or within lines of bamboos. In Jabalpur, Madhya Pradesh seedlings of B. bambos, B. nutans and D. strictus, were successfully intercropped with either maize or soya bean. In Thailand, the bamboo species are also intercropped with maize and peanut. In Sikkim, farmers grow D. hamiltonii and D. sikkimensis in agriculture fields all along the irrigation channels and stream banks to meet the fodder needs of their livestock. B. vulgaris and B. nutans have been grown on homesteads throughout Bangladesh. B. arundinacea is planted by farmersin depressed and water logged sites in Andhra Pradesh. Seshadri concluded that cultivation of soya bean (Glycine max) along with D. strictus was technically feasible and economically viable. The period of intercropping can be extended by adopting wider spacing of the bamboo culms and judicious manipulation of the bamboo canopy.

Bamboo + Conifer + Broadleaf Timber Trees

This model can be established by either converting semi-naturally mixed stands or planting new ones. The ratio of bamboos to trees is important, and in semi-naturally mixed stands this may be 7:3 or 8:2 for bamboos and broadleaf trees. The planting time for bamboos and trees should be determined based on the growth rate of the tree species involved. The tree species best adapted to bamboo crops are Albizzia sp., Gmelina arborea, Tectona grandis, Lagerstroemia parviflora, Anogeissus latifolia, Phyllanthus emblica, Zizyphus xylocarpa, Bombax ceiba, Stereospermum suaveolans, Melia azedarach, Aegle marmelos, Lannea grandis, Spondius pinnata, Erythrina indica, eucalypts, poplars, Dalbergia sissoo, etc., owing to their peculiar deciduous light crowns. Areca catechu with its narrow crown is also found to grow satisfactorily together with bamboo clumps. All species with deep umbrageous crowns like Adina cordifolia, Ficus spp., Mangifera indica, Artocarpus heterophyllus, Litchi chinensis, etc. should be avoided.

Bamboos + Tea Model

In this model, bamboos are often planted at spacing of 6 m x 4 m and tea plants at 2 m x 0.5 m. Intercropping of seasonal agriculture crops such as soya bean and vegetables can be done for one to three years after planting. However, care should be taken to leave enough space for unhindered growth of bamboos and tea plants.

Bamboo as Windbreak

Bamboos can be planted as windbreaks on the boundaries of agriculture field and orchards for protecting them from high speed wind. The practice is very common among the farmers in different part of the country. Mango orchards in tarai areas of Nepal are intercropped with agriculture crops, and the boundaries of orchards are planted with one or two rows of D. sissoo and D. strictus. In Cooch Behar, Dinajpur, Haldibari, Mayanaguri and Jalpaiguri areas in India, the clumps of B. balcooa, B. bambos and B. nutans have been cultivated in close spacing in one to two rows along the north western sides of rice fields as windbreaks against the dry and cold wind blowing from Nepal and Bihar.

Bamboo + Medicinal Plants

The model is suitable for hilly areas in the sub-tropical monsoon climatic zone that has a mild climate and adequate rainfall. The medicinal plants are chosen to suit the topography of the site. Shade loving plants of family Araceae and Zingiberaceae can be selected for intercropping as per the choice of demand and market. Species like Aloe vera, Catharanthus roseus, Cassia angustifolia, Curcuma domestica, Plantago ovata, Withania somnifera, etc. can tolerate shade and are suitable to grow as intercrop due to wide adaptability.

Bamboo + Crops + Fish Ponds

This model is usually made on the plain and wetlands where fish ponds are built. One to three rows of shoot-producing sympodial bamboos may be planted on the banks of pond, and crops such as soya bean and rye intercropped between bamboo clumps to form a complete food chain. Crops can be harvested as food of fish feed. Bottom mud from the ponds may be dug out in winter and used as fertilizer for bamboo clumps. The clumps of major cultivating bamboo species (B. balcooa, B. mabbosa, B. nutans, B. vulgaris and D. strictus, etc.) of south Asian countries may be replanted at 20-25 years of age for improving the productivity.

Bamboo + Edible Fungi

There are a large number of edible fungi (Dictyophora tomentosa, Planrotus ostreatus and Auricularia auricula-judoe) regarded as natural food rich in vegetable proteins grown in south China. Inoculation of Dictyophora sp. is done in September for varieties that grow in normal temperature and in May-June for those that require a higher temperature. Harvest is after four to eight months, depending on the fungus variety. P. ostreatus is inoculated in March and harvested two months later. A. auricula-judoe needs to be cultured in bags filled with the growth medium and hung on the bamboo.

Bamboo + Poultry/Dairy Farm

Bamboos can be used to ameliorate problems associated with the containment of animal wastes, nitrate concentration and slurry storages during dairy, beef and poultry operations. Bamboos can tolerate enormous N sources and turn it into marketable biomass. Bamboos-chicken agroforestry system is a new and common pattern in the hilly regions of Southern China and has high potential for extension throughout China. Soil nutrients and earthworm dynamics under this system have been evaluated and found that soil nutrients were improved but soil organism indicators were more sensitive than chemical ones. Earthworm quantity and mass between bamboos-chicken system and only bamboo forest were significant. While exploring a number of forage crops that hold the potential ofsupplying forage during lean period for dairy cattle and goats, bamboos have become a prime candidate as a perennial forage species making dormant seasons harvest possible. Having high protein content (12-19%), bamboos are comparable to alfalfa in nutritional value yet, they do not require intensive cutting, drying and storage processes of an annual crop.

AGROFORESTRY SYSTEMS AND ENVIRONMENTAL QUALITY

Agroforestry is perhaps as old as agriculture itself. The practice of growing trees and crops on the same unit of land has been prevalent for many centuries in different parts of the world, especially under subsistence farming conditions. Only during the past three decades, however, have these traditional integrated practices been brought under the realm of modern, scientific land-use scenarios. Modern agroforestry started primarily in the tropics, where the inherent advantages of age-old integrated farming systems began to be appreciated when the escalating land-management problems such as tropical deforestation, soil degradation, and biodiversity decline became serious concerns. Agroforestry began to be recognized and incorporated into national research and development agendas in many developing countries during the 1980s and 1990s. In the temperate, industrialized regions, too, the use of trees on agricultural landscapes for a variety of objectives has been an old practice; however, "modern" agroforestry in these regions had perhaps a slower evolution than in the tropics. It started with the general public's understanding about the environmental consequences of high-input land-use systems such as invasion of exotic species and alterations to nutrient, energy, and water flows leading to soil erosion, water-quality deterioration, and environmental pollution. Agroforestry gradually gained acceptance as an appropriate approach to addressing some of these environmental problems, and agroforestry initiatives started in North America, China, Australia and New Zealand, and southern Europe.

The concept of agroforestry is based on the expected role of on-farm and off-farm tree production in supporting sustainable land-use and natural-resource management. While the aboveground and belowground diversity provides more stability and resilience for the system at the site level, the system provides connectivity with forests and other landscape features at the landscape and watershed levels. These ecological foundations of agroforestry systems manifest themselves in providing environmental services such as soil conservation, carbon storage, biodiversity conservation, and enhancement of water quality.

In general, in the tropics where small family farms, subsistence food crops, and declining soil productivity are common features, the emphasis on the role of trees in improving soil quality of agricultural lands and providing food and nutritional security are characteristic of the numerous and diverse agroforestry systems. On the other hand, environmental protection is the cornerstone of agroforestry motivations and applications in the industrialized nations; it needs to be noted however, that although there is a tendency to equate "temperate" with "industrialized" regions, some temperate countries (or temperate regions of countries) such as China, Russia, Chile, and Argentina may be less motivated by environmental concerns than Canada, the United States, Europe, and New Zealand and Australia. Because of these differences in the role of agroforestry in the tropics and temperate (industrialized) regions, the expectations of environmental benefits of agroforestry systems (AFS) have had slightly different connotations in the two contexts. In the tropics, soil-related issues such as soil fertility improvement, soil and water conservation, and reclamation of saline-, acid-, and other degraded soils were the major areas of agroforestry research and application during the 1980s and 1990s. Although these are environmental quality issues, they were pursued mainly as a means of augmenting the soil's capacity to produce crops sustainably. In the temperate regions, on the other hand, agroforestry was seen more as an environmental protection opportunity through exploiting the beneficial role of trees for water-quality enhancement in agricultural watersheds, reduced wind and water erosion, and so on. In both situations, however, the focus was at the local or regional scale. During the past decade, with increasing realization of the global impact of human activities on the environment, environmental issues of a global scale such as carbon sequestration/climate change and biodiversity conservation have become increasingly important. Today, the environmental-quality and ecosystem-service functions that are commonly associated with agroforestry are C sequestration/climate change, biodiversity conservation, water-quality enhancement, and soil improvement and these benefits transcend from local to global levels according to the "act locally, think globally" dictum. The subject matter of this special collection of papers is, therefore, timely and appropriate.

Water Quality

Studies on tropical AFS in nutrient-depleted soils have demonstrated the ability of deep-rooted trees to absorb nutrients that have leached below the rooting zone of agronomic crops and to recycle these nutrients via leaf litter and fine root turnover, thus improving the nutrient-use efficiency in the system as a whole. Similar results have also been reported from studies on temperate agroforestry systems. This so-called safety-net effect of tree roots could have an important application in the heavily fertilized sandy soils that have low nutrient retention capacities. The capacity of tree roots to capture nutrients from the deeper soil horizons can enhance nutrient storage in the plant–soil system and thereby reduce the amount of nutrients that might otherwise be transported to ground and surface water through runoff and leaching, thus causing nonpoint-source pollution of these water bodies. Schultz et al. described the water-quality enhancement benefits of riparian buffer systems (which is one of the five principal agroforestry practices in North America, the others being alley cropping, forest farming, silvopasture, and windbreaks;). Reviewing the results from around the world, they concluded that in the right location, riparian forest buffers can remove significant amounts of sediment, nutrients, and pesticides from both surface and subsurface waters.

Carbon Sequestration and Climate Change Mitigation

Agroforestry systems are believed to have a higher potential to sequester C because of their perceived ability for greater capture and utilization of growth resources (light, nutrients, and water) than in single-species crop or pasture systems. The estimates of C stored in AFS range from 0.29 to 15.21 Mg ha^{-1} yr^{-1} aboveground and 30 to 300 Mg C ha^{-1} up to 1 m depth in the soil. Recent studies under various AFS in diverse ecological conditions showed that tree-based agricultural systems, compared to treeless systems, stored more C in deeper soil layers near the tree than away from the tree, and higher soil organic C content was associated with higher species richness and tree density. Furthermore, C3 plants (trees) contributed to more C in the silt + clay (<53 μm diameter) fractions that constitute more stable C than C4 plants, in deeper soil profiles. The amount of C sequestered in AFS depended to a great extent on environmental conditions and system management.

Soil Improvement

One of the tree-mediated benefits of considerable advantage in the tropics is that trees and other vegetation improve the productivity of the soil beneath them. Research results during the past two decades show that three main tree-mediated processes determine the extent and rate of soil improvement in agroforestry systems. These are (i) increased nitrogen input by N2–fixing trees, (ii) enhanced availability of nutrients resulting from production and decomposition of tree biomass, and (iii) greater uptake and utilization of nutrients from deeper layers of soils by deep-rooted trees. Furthermore, the presence of deep-rooted trees in the system can contribute to improved soil physical conditions and higher soil microbiological activities under agroforestry. Farmlands in many parts of the developing world generally suffer from the continuous depletion of nutrients as farmers harvest without fertilizing adequately or fallowing the land. One promising way for overcoming the acute problem of low nutrient status of soils, such as African soils in general, is to enable smallholders to use fertilizer tree systems that increase on-farm food production. The other major avenue of soil improvement with agroforestry is through soil conservation. About 1.9 billion ha of land, a third of total farmland, in developing nations are estimated to be degraded through erosion, salinity, and fertility depletion. The potential of agroforestry to reduce the hazards of erosion and desertification as well as to rehabilitate such degraded land and to conserve soil and water has been well recognized. The soil ameliorative potential of agroforestry systems has recently been demonstrated in the temperate zone as well.

Biodiversity Conservation

When landscapes are increasingly being fragmented and remaining patches of natural vegetation are reduced to isolated habitat islands consequent to human activities, mixed-species AFS could play a significant role in helping to maintain a higher level of biodiversity and provide greater landscape connectivity (Montagnini et al., 2011). This can occur in at least three ways: (i) intensification of AFS leading to reduced exploitation of protected areas, (ii) increasing biodiversity in working landscapes through the expansion of AFS into traditional farmlands, and (iii) increasing the species diversity of trees in farming systems. Where croplands occupy most of the landscape, riparian forest buffers and field shelterbelts can be essential for maintaining plant and animal biodiversity, especially under a changing climate scenario. Trade-offs between ecosystem conservation and agricultural production can convincingly be addressed by shifting the view from the plot to the landscape scale and integrating biodiversity-friendly land-use systems such as agroforestry into development strategies.

Some Methodological Challenges

Methodological problems are common to most land-use systems, but they are of a higher order of magnitude in AFS compared with agricultural systems because of the integrated nature of AFS and the relative newness and slow growth rate of research on the subject. The methods and procedures used and estimations made in agroforestry research vary widely; therefore, the available datasets are mostly inconsistent and often incomparable and inconclusive. Some of the common problems include the following:

- Allometric equations: Aboveground biomass in AFS and forestry systems is often estimated based on allometric equations. But rigorous allometric equations are not available for many tree species used in AFS, especially in the tropics. This leads to broad generalizations and inaccuracies.
- Soil sampling depth: Most soil studies are limited to surface soils down to 20 or 30 cm depth. Sampling beyond the surface soil is important while studying tree-based systems such as agroforestry, not only because tree roots extend to deeper soil horizons but also because of the role of subsoil in long-term nutrient dynamics and C stabilization.
- Pseudoreplication: As in ecological and forestry studies, it is common in agroforestry research to use preexisting field plots of the same contiguous experimental unit without true replicates of treatments. Statistical comparisons between treatments of such studies may be invalid. Some argue that in studies on large plots, spatial interspersion of replications together with the use of a systematic design can be used to alleviate possible pseudo-replication problems. Forestry researchers have used composite samples drawn from large experimental units as replicates considering the land-use systems as fixed-effect treatments. In the statistical sense, a fixed-effect model means that inferences are restricted to the treatments in the study; the results cannot be used to make conclusions about other agroforestry systems.
- Repeated measures: The fixed-effects model also applies to the so-called repeated measures, which refer to measurements made in time or space on the same subject or experimental unit, such as a tree or a plot. For example, in agroforestry experiments, we may have to draw soil samples from depth increments from the same sites, or transects at defined horizontal distances from trees. When time and space are considered as treatments, they cannot be randomly assigned; the depth/distance increments are treated as repeated measures rather than as independent measurements. The nonrandomized nature of repeated measures designs often results in the violation of the assumptions necessary for valid univariate analysis. However, statistical procedures are available to address the limitations imposed by the model. In the case of soil depths at the same site, they could be stratified and each soil depth considered independently treating each site as a replication.
- Chronosequence studies: Studies that are done to understand the change in soil C and nutrients are often based on the assumption that the changes in C and nutrient stocks are likely to be linear through time. This may not necessarily be

true. In addition, in the case of C, it is difficult to know if the residence time of C that is sequestered initially in a system differs from that of later-sequestered C. Several such challenges need to be addressed for realistically assessing the impact of agroforestry practices.

References

- Modelling-agroforestry-systems-333696631: researchgate.net, Retrieved 16 June, 2020
- Medicinal-and-aromatic-plants-in-agroforestry-systems-226462494: researchgate.net, Retrieved 28 May, 2020
- Lalnunmawia, F. 2003. Agroforestry with bamboo and ginger to rehabilitate degraded areasin-NorthEastIndia. JournalofBambooandRattan,2(2):103-109
- Overstory-35-animals-in-agroforestry, the-overstory-229: agroforestry.org, Retrieved 22 July, 2020
- Benefit-cost analysis of bamboo in comparison with other crops in mixed cropping home gardens in Kerala state, India. 2004. Journal of Bamboo and Rattan, 3(2): 99-106

PERMISSIONS

All chapters in this book are published with permission under the Creative Commons Attribution Share Alike License or equivalent. Every chapter published in this book has been scrutinized by our experts. Their significance has been extensively debated. The topics covered herein carry significant information for a comprehensive understanding. They may even be implemented as practical applications or may be referred to as a beginning point for further studies.

We would like to thank the editorial team for lending their expertise to make the book truly unique. They have played a crucial role in the development of this book. Without their invaluable contributions this book wouldn't have been possible. They have made vital efforts to compile up to date information on the varied aspects of this subject to make this book a valuable addition to the collection of many professionals and students.

This book was conceptualized with the vision of imparting up-to-date and integrated information in this field. To ensure the same, a matchless editorial board was set up. Every individual on the board went through rigorous rounds of assessment to prove their worth. After which they invested a large part of their time researching and compiling the most relevant data for our readers.

The editorial board has been involved in producing this book since its inception. They have spent rigorous hours researching and exploring the diverse topics which have resulted in the successful publishing of this book. They have passed on their knowledge of decades through this book. To expedite this challenging task, the publisher supported the team at every step. A small team of assistant editors was also appointed to further simplify the editing procedure and attain best results for the readers.

Apart from the editorial board, the designing team has also invested a significant amount of their time in understanding the subject and creating the most relevant covers. They scrutinized every image to scout for the most suitable representation of the subject and create an appropriate cover for the book.

The publishing team has been an ardent support to the editorial, designing and production team. Their endless efforts to recruit the best for this project, has resulted in the accomplishment of this book. They are a veteran in the field of academics and their pool of knowledge is as vast as their experience in printing. Their expertise and guidance has proved useful at every step. Their uncompromising quality standards have made this book an exceptional effort. Their encouragement from time to time has been an inspiration for everyone.

The publisher and the editorial board hope that this book will prove to be a valuable piece of knowledge for students, practitioners and scholars across the globe.

INDEX

CPSIA information can be obtained
at www.ICGtesting.com
Printed in the USA
LVHW061754140222
711000LV00009B/61

9 781639 890361